T0100685

THE MAJORANA CASE

Letters, Documents, Testimonies

THE MAJORANA CASE

Letters, Documents, Testimonies

Erasmo Recami

University of Bergamo, Italy
INFN-Sezione di Milano, Italy
UNICAMP, Campinas, SP, Brazil

NEW JERSEY • LONDON • SINGAPORE • BEIJING • SHANGHAI • HONG KONG • TAIPEI • CHENNAI • TOKYO

Published by

World Scientific Publishing Co. Pte. Ltd.
5 Toh Tuck Link, Singapore 596224
USA office: 27 Warren Street, Suite 401-402, Hackensack, NJ 07601
UK office: 57 Shelton Street, Covent Garden, London WC2H 9HE

British Library Cataloguing-in-Publication Data
A catalogue record for this book is available from the British Library.

THE MAJORANA CASE
Letters, Documents, Testimonies

ISBN 978-981-120-701-3
ISBN 978-981-120-769-3 (pbk)

For any available supplementary material, please visit
https://www.worldscientific.com/worldscibooks/10.1142/11465#t=suppl

Desk Editor: Rhaimie Wahap

Typeset by Stallion Press
Email: enquiries@stallionpress.com

Printed in Singapore

Preface

Ettore Majorana was a theoretical physicist of the group founded in Rome by Enrico Fermi, now known as the "Second Prometheus" since in Chicago in 1942 he constructed the first nuclear reactor, producing peaceful energy one million times greater than the one obtainable by combustion or other chemical reactions. Enrico Fermi regarded Majorana "a genius like Galileo and Newton": something that many scholars in the world are beginning to recognize, at least in the last decade, especially in consideration of the results presented by him in his last paper of 1937. Yes, first appeared in 1937: and fully recognized 70 or 80 years after… In that article, initially known for the important introduction of the Majorana neutrino, it has been noted, much later, the introduction of the Majorana fermions (detected experimentally during the last 12 years in condensed state physics; and shown to be an essential stepping stone for building up quantum computers!), and also of the Majorana algebras (soon found to be essential, for example, in the study of properties of the mathematical so-called *Monster Group*).

The author moved in 1968 from the Milan University's Physics Department to the one of the University of

Catania (the home-town of Ettore Majorana), and soon started to investigate Ettore Majorana's life and work, in full cooperation and harmony with the Majorana family of Catania — mainly Ettore's sister-in-law and his sons, Ettore's nephews — and of Rome: mainly Maria Majorana, Ettore's wonderful sister. In a few years, the author discovered Ettore's correspondence and began collecting the large majority of the serious documents still in existence on Majorana's life and work (and being the first to publish all of them), with the help of Ettore's former friends and schoolmates, and of a few colleagues. The author then became the biographer of such a genius (the author win for instance the 2000 Prize for History of Physics from SIF, the Italian Physical Society). In 1972, he passed some documents — besides to Edoardo Amaldi — to the famous Italian writer Leonardo Sciascia, who used them for writing down a well-known essay on Majorana's disappearance. From the scientific point of view, the author was fortunate in meeting and getting the help of some interested fiends, in particular S. Esposito.

Earlier we mentioned Majorana's disappearance. Indeed, he mysteriously disappeared at 31, in 1938; and that mystery has remained unsolved. Majorana's life is fascinating not only for his personal and scientific stature, but not less for his disappearance which makes his biography akin to a detective story.

For readers who are interested to know more, the writer may be contacted at recami@mi.infn.it. His homepage is www.unibg.it/recami. It could be interesting for the readers to know that, following the initial (1987) Italian edition of this book, two documentary films by Italian broadcasters were released in 1987 and 1990. They are still deemed interesting even though they are in Italian. The two documentaries may be viewed at https://www.youtube.com/watch?v=RLqHu2w7Rds and https://www.youtube.com/watch?v=bv4DlzOTLKg. Two more videos are indicated in the Bibliography.

Acknowledgments

This translated version would not exist without the generous contribution of Carmelo Papa (former Executive Vice President of STMicroelectronics), Sean M. Donohoo, Nunzio Abbate, and others (including Andrea Lastrina), all from STMicroelectronics, Catania. We are deeply indebted to all them for their strong interest and effort. Let us also recall the preliminary work by Sara Keller, performed long ago with a financial aid offered by Dharam V. Ahluwalia.

We willingly acknowledge World Scientific Publishing Company for the interest in publishing this book and its editor, Rhaimie Wahap.

The original Italian version would not have existed without the initial support of the Majorana family both in Rome and Catania (in particular Maria, Ettore Majorana's sister; Fabio, Ettore Jr. and Pietro, Ettore's nephews, and of their mother Nunni Cirino), of George Sudarshan, Eduardo Caianiello, Marco Mondadori, Gianni E. Ferrari, Ferruccio Parazzoli, Maria Agrati, Libera R. Baldini, Ioannes M. Besieris, and Kirk McDonald. Nor without the former, friendly help by Sante Di Renzo and his Editorial Office.

Letters, documents, testimonies have been personally discovered or collected by us during the last five decades, except for the following material: (i) that from the Archivio Generale dello Stato, Rome, which was copied to us by Donatello and Fosco Dubini, of Cologne's "Filmproduktion", with whom we operated a proficuous information exchange; (ii) the letters by Ettore M. to Giovanni Gentile Jr. (the son of a well-known Italian philosopher and statesman), which reached us partly as a gift of Benedetto and Enrico Gentile, and partly due to the interest of Leonardo Sciascia; (iii) the letters sent to Ettore M. between 1933 and 1937 by his uncle Quirino Majorana — a talented experimental physicist — (which will be reported here only in small part, because of their technical character), that were given to us in 1988 by Silvia Majorana Toniolo through Franco Bassani's courtesy; (iv) the correspondences exchanged in 1933 by Ettore M., then at Leipzig, with the CNR (National Research Council), forwarded to us by Michelangelo De Maria with Gigliola Fioravanti's collaboration. To them all we are obliged: as well as to Bruno Russo who, in producing a meticulous TV film, further recorded highly interesting testimonies.

Moreover, let us thank Edoardo Amaldi (and SIF) for having allowed us in due time to add, as an Appendix, his *Ricordo di Ettore Majorana*, besides some quotations in the text; and for this last reason, let us thank also Mario Bunge, Bruno Pontecorvo, Leonardo Sciascia, and Emilio Segré (and the publishers Boringhieri, Editori Riuniti, Einaudi, and Zanichelli). Analogously, we thank editorships and publishing houses of *Giornale critico della filosofia italiana* and of *Storia contemporanea.* Quite useful to this publication are also the authorizations, from their authors, to reproduce the many letters and testimonies that appear in this book. We are grateful in particular to Daniele Amati, Blanca de Mora Asturias, Michelle Bergadaa, Gilberto Bernardini, Hans Bethe, Giuseppe Cocconi, Aurora Fornoni Bernardini, Yuval Neeman, Rudolf Peierls, Eliano Pessa, Gastone Piquè, Bruno Pontecorvo, Franco Rasetti, Tullio Regge, Carlos Rivera, Gianni Sansoni, Angelo Savini-Nicci, Leonardo Sciascia, Emilio Segrè, Carla Tolomeo, Giancarlo Vigorelli, Enrico Volterra, Gleb Wataghin, Victor Weisskopf, and GianCarlo Wick. We cannot forget, moreover, the help and the stimuli received for 30 years by Dharam Ahluwalia, Nello Ajello, Francesco Alberoni, Edardo Amaldi, Marcello Baldo, Franco Bassani, Luiz Bassani, Roberto Battiston, Carlo Becchi, Gilberto Bernardini, Piero Bianucci, Quirino Bortolato, Carlo Castagnoli, Alberto Castoldi, Mario Ciancio, Flora Cirino, Santi Correnti, Aldo Covello, Richard H. Dalitz, Ubiratan D'Ambrosio, Carlo Dignola, Calcedonio Di Salvo, Federico Di Trocchio, Viviano Domenici, Antonino Drago, Giorgio Dragoni, Salvatore Esposito, Myron Evans, Amand Faessler, Roberto A. Ferrari, Maristella Fracastoro, Alberto Gabriele, Guido Gerosa, Mario Giambiagi, Enrico Giannetto, Adriano Gozzini, Giancarlo Graziosi, Françoise Gueret, Philippe Gueret, Remo Guerrini, Hugo Hernández Figueroa, Josè Leite-Lopez, Arrigo Levi, Dominique Leglu, Aldo

Magnano, Aurelio Manzoni, Dora Maiolo, Angelo Majorana, Pietro Majorana, Salvatore T. Majorana, Simão Mathias, Roberto Mignani, Remo Morzenti Pellegrini, Enrico Nistri, Giueppe Occhialini, Paolo Palazzi, Stefano Paleari, Raffaele Pisano, Salvo Ponz de Leon, Franco Prattico, Bruno Preziosi, Madeleine Rasetti Hennin, Renato A. Ricci, Vanni Ronsilvalle, Remo Ruffini, Bruno Russo, Salvatore Scalia, Gilda Senatore, Paolo Strolin, Franco Strumia, Pasquale Tucci, Ettore Vaccaro, Enzo Vitale, John N. Warnow, Bruce R. Wheaton, Sir Denys Wilkinson, Michel Zamboni Rached; and by the friends (Segnini, De Renzini, Maccagni, Colotto, Ronchi, Tricarico, Puccianti e Guerri) of "Domus Galilaeana"; by SIF; and especially by my wife Marisa T. Vasconselos. Lastly, for the reading of the original manuscript, thanks are willingly due to Piero Dell'Acqua, Maria Majorana, Roberto Mignani, Umberto Recami, Bettina Roberto, Laura R. Sansoni, and Giancarlo Vigorelli; while for the English version, thanks are due also to Mario S. Casella.

The author is presently visiting Decom/Feec of UNICAMP (SP, Brazil) — on leave from the Italian Bergamo University and INFN-Milano — due to a PVE fellowship of the Brazilian FAPESP.

E. R.

World Scientific Publishing Company would like to thank the Majorana family and Prof. Erasmo Recami for the use of their personal collection of photos in this volume.

Contents

Part I

Ettore Majorana: The Man and the Scientist

1

The Disappearance

I've made a decision that was by then inevitable. There is no grain of selfishness in it
ETTORE MAJORANA

The Last Letters of Ettore Majorana

Born in Catania, Italy, on the 5th of August 1906, Ettore Majorana was granted a professorship in theoretical physics "due to a high reputation of exceptional ability" at the University of Naples on November 1937. On Friday, 25th March 1938, the 31-year-old Majorana sent a letter to the institute director Prof. Antonio Carrelli, who received it at 2 o'clock the following day. The letter read:

> "Naples, 25th March 1938-XVI — Dear Carrelli, I've made a decision that was by then inevitable. There is no grain of selfishness in it, although I recognize the bother that my sudden disappearance will bring to you and the students. I ask your forgiveness also for this, but above all for having betrayed the trust, sincere friendship, and kindness that you have demonstrated to me over these months. I ask you also to give my best to all who I came to know and respect at your institute, Sciuti in particular, and for whom I will hold dear memories,

at least until eleven o'clock this evening, and possibly thereafter — E. Majorana."

On the table of his room at the Bologna Hotel in Via Depretis street in Naples, he left an envelope on which was written *To my family*, and inside, these few lines:

"Naples, 25th March 1938-XVI — I have just one wish: that you do not wear black. If you wish to mourn me then do so, but not for more than three days.[1] Afterward, if you can, keep my memory in your hearts, and forgive me — affectionately, Ettore."

Then, it appears, he placed his passport in his pocket and withdrew the money he had earned in his first three (or four) and a half months of teaching at the university, and boarded the steamship "Postale" of the Tirrenia company, which provided ferry services between Naples and Palermo. The ferry left port at 10:30 pm. All this seems to indicate that he intended to end his life, or in any case to disappear.

Instead, the following day, a Saturday, from Palermo he immediately sent Carelli an urgent telegram in which he retracted the letter he had previously sent from Naples, and took a room at the grand hotel "Sole" in Corso Vittorio Emanuele. It was on this hotel's letterhead that

---

[1] The customary three days of "deep mourning" in the Sicilian tradition.

he wrote a second letter to Carrelli, which represents his last remaining signed document:

> "Palermo, 26th March 1938-XVI — Dear Carrelli, I hope that the telegram and the letter arrived together. The sea has refused me and tomorrow I will return to the Bologna Hotel [*in Naples*], perhaps together with this letter.[2] I do, however, intend to renounce my teaching post. Do not take me for an "Ibsenian girl"[3] because the case is different. I am at your disposal for additional details — affectionately, E. Majorana."

The evening of the same Saturday, the Postale departed from Palermo bound for Naples, where it was expected to arrive at 5:45 Sunday morning. Majorana purchased a seat in a passenger compartment. All the evidence leads to the presumption that Ettore intended to return to Naples. Instead, either during the voyage or immediately thereafter (or possibly just before), he disappeared.

From the State Central Archives

Regarding the events leading to his disappearance, we have several documents and witness testimonies.

---

[2] The Palermo–Naples steamer also provided postal services.

[3] A reference to a character in the work of Norwegian dramatist Henrik Ibsen.

Coming from the Ministry of the Interior (and precisely by its Division of General and Classified Affairs), these documents are today stored at the State Central Archive[4] in Rome. A note written on the letterhead of the Council of State, evidently on 31st March,[5] reads:

> "Prof. Ettore Majorana...: I) On 25-March (Friday) wrote from Naples a letter to the Director of the Institute of Physics, Prof. Carrelli, stating he had to take the inevitable decision of abandoning his teaching. He informed him that he was departing from Naples by sea [*from Naples it is easy to reach Sicily by sea*]. He left the Bologna Hotel, where he had been staying, at around 5 p.m.; II) 26-March (Saturday) from Palermo, he informed Prof. Carrelli by letter that he was returning, perhaps together with the same letter, to Naples, where he should have arrived on Sunday the 27th or Monday the 28th. The same day (Saturday) from Palermo he telegraphed the hotel requesting that they keep his room, where he had left clothing and papers; III) 31-March (Thursday). Since that morning, there has been no additional news of him. Prof. Carrelli reported his disappearance to the Police commissioner of Naples; IV) The family anxiously searched for him. It turned out that no one by his name had taken the Naples-Palermo/Palermo-Naples steamship [*news which was subsequently corrected by Ettore's family*]; V) The young man, a severe misanthropist in

[4] Series PS — 1939-A1, envelope 51.

[5] Registered with no. 5691-A1 on the date of 1st April 1838.

poor health,[6] could be in hiding somewhere in Palermo, or at a private clinic. One could suppose that he has gone to Tunisia [*if he wanted to go abroad, from Palermo it is easy to reach Tunisia by sea*]. It's improbable that he went to a place where he was well-known (to Catania, for example); VI) He has a passport for Europe, renewed last June or July; VII) Height: 1.68 m. Elongated face; eyes: large and bright; black hair, brown skin. Iron-gray overcoat; dark brown hat."

Following this information, transmitted to Rome most probably by the Police Commissioner of Naples by telephone, the Chief of Police, Senator Arturo Bocchini, addressed to all the Commissioners of the Italian Kingdom the telegram[7]: "442 Stop Please investigate for the purpose of tracing, without his being aware of it, the University of Naples physics professor Ettore Majorana, son of the late Fabio and Dorina Corso, born in Catania 5th August 1906, who left Naples without providing any news to family Stop Professor Majorana has passport for European States renewed June or July last year Stop Description Stop Height 1.68 Elongated face Large dark eyes Black hair Dark complexion Dressed in iron gray overcoat Dark

---

[6] On 20th January 1938 Ettore himself, when filling in the "matriculation status" form at the University of Naples, wrote: "HEALTH: *somewhat delicate*".

[7] Transmitted to the "Cipher Office" the same day, 31st March, with no. 10639.

brown hat Stop If found telegraph urgently Ministry reporting any movement and location where heading Stop Chief of Police". The text of the telegraph dispatch, however, bears the successive annotations: "No reply: 7/4/38"; the stamp "Evidence expired", and then "To be filed": the order of archiving.

In the meantime, on Wednesday, 30th March, Antonio Carrelli had turned to the university Rector with a letter labeled "Strictly confidential":

"*Magnifico Rettore,*
it is with great sorrow that I communicate to you the following:

On Saturday, March 26th at 11 in the morning I received an urgent telegram from my colleague and friend Ettore Majorana, professor of theoretical physics at this University, conceived in these terms: "do not be alarmed. A letter will follow. Majorana". This message was incomprehensible to me. I inquired and found that on that morning he had not delivered his lecture. The telegram came from Palermo.

With the 2 p.m. mail delivery, I received a letter from Naples dated the previous day, in which he expressed suicidal intentions. I understood then that the urgent telegram from Palermo the following day was in fact intended to reassure me, providing me proof that nothing had happened. In fact, Sunday morning I received a letter by express mail from Palermo, in which he told me that these bad feelings had disappeared and that he would return soon.

Unfortunately, he did not appear the next day, Monday, neither at the Institute nor at the hotel where he had been staying. A bit alarmed at his absence, I passed the news of what had happened to those of his family who were residing in Rome. Yesterday morning, his brother [*Salvatore*] visited me, and he and I went to the Police Commissioner of Naples, requesting that he inquire with the police headquarters in Palermo as to whether Prof. Majorana was still staying at a hotel in that city. Since, as of this morning, I have still not received any news, I inform you of what has occurred in the hope that my colleague has simply sought some respite following a period of exhaustion, of unhappiness, and will soon be back among us to offer his important contributions of work and intellect."

With deference,
Antonio Carrelli

From this statement we learn, and not without astonishment for those familiar with Italy's postal system today, that express letters and telegrams arrived on time, even on Sundays. And Majorana will have certainly given this due consideration. Moreover, from the notes of the State Councillor — perhaps Bocchini himself — we learn that Ettore had left the Hotel Bologna at about five o'clock in the afternoon. One might then ask how he spent the hours prior to boarding the steamship. Did he walk around? Did he have dinner? He almost certainly already had the Naples-Palermo ticket in his pocket. Regarding the tickets for the departure and return voyages, it should be mentioned that later on the ferry company Tirrenia declared to the family that they had tracked them both. To the extent that Ettore's brother, Salvatore, procured a letter from Senator Giovanni Gentile (the renowned philosopher), and sought a meeting with the Chief of Police — we can imagine with what degree of anxiety — to inform him of the new findings: "Prof. Majorana ... it was believed that he remained in Palermo. However, this hypothesis is now discredited by the fact that the return ticket was found by the management of Tirrenia, and because he was seen at five o'clock (*in the morning*) in a passenger compartment of the steamship during the return trip — still asleep. Then, in early April, he was seen — and recognized — at Naples, between the Palazzo Reale and the Galleria while walking up from Santa Lucia, by a nurse who knew him and who also guessed the color of his overcoat...", where *guessed* implies that the nurse, when questioned by the family, was able to provide the correct color of his overcoat: iron-gray. The previous passage is from another file, in the State Central Archive: series "Political Police: Personal", envelope no. 780; from which we learn that the meeting of Salvatore with Arturo Bocchini took place on 18th April. But we will not dwell on this, given that this file had already been found and revealed by the Italian writer Leonardo Sciascia.[8]

Testimonies from the Period

"He was seen at five a.m. in the passenger compartment of the ship during the return trip, and was still asleep". Seen by whom? Probably by Vittorio Strazzeri, professor of geometry at the University of Palermo who, according to Tirrenia's records, had traveled in the same three-bed compartment and from whom we have a written account, although this was taken a few months later. Writing to Salvatore on 31st May 1938, Prof. Strazzeri said:

> "Dear Mr. Majorana, It is my firm belief that, if the person who traveled with me was your brother, he did not dispatch himself, at least not before our arrival in Naples. This is because, when I got up [*from the bed*], we were in front of the port of Naples and, being a *very clear day*, many passengers were on the deck of the ship. I repeat that I did not see any luggage in the compartment, but what caught my attention was that his vest, or perhaps it was a jacket (in any case, some clothing), had been laid on the netting that covers each bed. This

[8]Because of our old links, resulting in a friendship since the early 1970s, the author was called 30 years later to serve as the president of the Associazione "Amici di Sciascia" ("Friends of Sciascia" association) for a two-year period.

caught my attention because my biggest concern when traveling is the safekeeping of my wallet and passport. I do not question that the third passenger was called *Carlo Price*, but I can assure you that he spoke Italian like *we from the South of Italy do*, and also that he seemed to me to be some kind of shopkeeper, or of lower status, as he lacked that unconscious refinement which comes from high culture... I repeat: if the young man who traveled with me was your brother (I say "young" because his hair was full, and because he gave me that general impression), he certainly did not kill himself before the ship's arrival in Naples. Please kiss your mother's hands for me, and give my regards to your family. If you have any news, please inform me. Be certain that if it is good news, as I hope and believe, it will bring me great joy. Yours sincerely, Strazzeri — Palermo, 05.31.1938 — *P.S.*: Forgive me if I dare to offer you a suggestion, which is to explore the possibility that your brother has closed himself up in a convent, as has happened even with people who were not very religious, I believe at Monte Cassino."

The suggestion made by Strazzeri in the *Post Scriptum* is supported by another account (reported also by Ettore's mother, as we will see): the Jesuit priest De Francesco, in the photos shown to him by Ettore's family, recognized the distinguished young man who in late March or early April had appeared, rather troubled, before the Superior of the Church called Gesù Nuovo in Naples, requesting to be hosted on a retreat to experiment with religious life. However, when informed of the bureaucratic obstacles that this would entail, the young man thanked him, apologized, and left.

This statement by the priest, a former Jesuit Provincial, is supported by the fact that the Hotel Bologna, the Institute of Physics (in Via Tari street), and the Convent of Gesù Nuovo are all along the same natural route, and are in close proximity to one another. We ourselves, starting from Via Tari, had to pass along Via Depretis and in front of the church-convent. And this was taken very much into account by the family.

It is curious, however, that no one took note of a report which the Police Commissioner of Naples quite some time earlier had sent to the university Rector, who had personally involved him "for any help with the investigation that you can provide": "As Your Excellency knows, in March Prof. Ettore Majorana left the Hotel Bologna in this city, indicating his suicidal intentions to Prof. Carrelli of the Institute of Experimental Physics at your distinguished University; intentions which he did not carry out. At the request of his brother, Dr. Salvatore, a search began and was subsequently intensified, but it has been thus far unsuccessful. It emerged only that the missing person, apparently on the 12th of this month, appeared at the *Convent of St. Pasquale di Portici* requesting admittance to that religious order; but, the request having been denied, he left for an unknown destination. The investigation continues with diligence and, if the results are favorable, we will inform Your Excellency". Wartime events destroyed most of the university archive, and the Police Headquarters of Naples customarily destroy nearly all their records every five years. But we know this bit because the Rector took care to write it down for the Ministry of National Education in a letter that later arrived under our eyes: "Confidential — Following my previous communications, I transcribe here below the notes from last April 29th, number 87,966, of the Police Commissioner of Naples ...".[9]

This report, ringing true, would be of the highest interest *if* true because it would demonstrate (along with the testimony of the nurse) that Ettore was alive and in Naples two weeks after his disappearance, even if his name did not appear, it seems, in the registers of any hotel in the area.

At least the family, once again, gave this report its due consideration. Indeed, it is possible that the Commissioner is simply referring to the information they obtained: In fact, Ettore's brother Luciano (occasionally accompanied by his mother) went so far as to knock on the doors of convents in and around Naples carrying

---

[9] This letter, received by the Directorate General of Higher Education on 5th May, bears the diagonal penned inscription *"To be filed" for now, return in 15 days,* and, later, the stamp *Discharged.*

a photograph of Ettore. Mrs. Nunni, Luciano's widow (also recently passed away), recalled that there were only two positive responses: one from an open convent, and another from a cloistered one.[10] From the latter, a comment was made to his mother which left some doubt: "But why are you looking for him, Madam? The important thing is that your son is happy". Incidentally, Luciano often observed that it is easier to find the body of a dead man than a living one, and over time he would speak less and less of Ettore, confiding, "If Ettore decided to leave, we should respect his decision and search for him no more".

That Ettore sought refuge in a monastery, at least temporarily, seems to have been established; less certain is whether his wish was crowned with success. The majority of the most "confidential" documents available till now seem to hint at a different evolution of Ettore's life. But, as with any *detective story* worthy of respect, this will be revealed in due course. In any case, in response to persistent rumors, the family decided to send a petition to the Pope, and it was Maria, Ettore's sister, who delivered to Pope Pacelli a parchment which promised to abstain from any interference while asking only if the Vatican knew whether Ettore was alive. The Majorana family never received a response (let alone given access to the Vatican archives). Why? Because no attention was

[10] San Pasquale di Portici at that time was a cloistered convent.

given to their petition, or because the Vatican was bound to secrecy?

Official Proceedings

> "The State Secretary Minister of National Education, … Considering that Prof. E. Majorana was absent from his office without justification for a period longer than ten days, and considering that, despite the search conducted, no information regarding the aforementioned professor has been obtained, DECREES: As of 25th March 1938-XVI, Prof. E. Majorana is dismissed from his post. Rome, December 6th, 1938-Year XVI — Giuseppe Bottai."

From the date of the decree, we must acknowledge that the authorities waited quite a long time for news of his whereabouts. Particularly if you compare this with the much shorter period needed for his appointment: on 25th October 1937, the selection Committee (chaired by Enrico Fermi) for the professorship in theoretical physics in which Ettore participated, proposed to the minister that Ettore's appointment be granted *out of competition* "due to high and well-deserved fame". On 2nd November, Minister Bottai issued the decree of appointment, and on 4th December the decree was recorded by the Auditors' Court ("Corte dei Conti"). Thus, the appointment was announced to Ettore by the ministry at his home in Viale Regina Margherita 37 in Rome. Indeed, before disappearing, Ettore would receive his salary for the first months of his service.

We shall refer later to some interesting events that occurred during the competition for the professorship in which Ettore Majorana participated. But now, will read the decree of appointment, as it reveals the amount of money Ettore was able to collect prior to his disappearance:

> "The Minister, etc.: ...Having read art. 8 of the Royal Decree of 20th June 1935-XIII, no. 1071; Having considered the opportunity etc... DECREES: From the date of 16th Nov. 1937-XVI, Prof. E. Majorana, due to the high fame of singular expertise he attained in the field of the studies on Theoretical physics, is appointed Full Professor of Theoretical Physics at the Faculty of Mathematical, Physical and Natural Sciences of the Royal University of Naples. From the same date, Prof. Majorana is placed at grade VI, group A, with the salary of Lira 22,000, plus Lira 7000 s.s.a. [=save further adjustments], respectively reduced to L. 19,872 and L. 6323 s.s.a. This decree shall be communicated to the Corte dei Conti [*Auditors' Court*] for registration. Rome, 2 Nov. 1937-XVI — The Minister: Bottai."

Following the registration and conferring of his appointment, which was sent to his residence in Rome, on 12th January 1938 Ettore would respond from Naples:

> "... In respectfully offering to the Minister the expression of my gratitude for the high distinction granted me, I wish to affirm that I will give all my energy to the Italian school and science, today in fortunate ascent towards the re-conquering of its ancient and glorious supremacy. Yours faithfully, — Ettore Majorana."

These words may seem rhetorical, and are certainly words "suited to the occasion", but true, in the sense that from the early 1930s, thanks to the involvement of Enrico Fermi and his group, the Italian physical sciences were indeed in rapid development. Ettore could well afford to assert his own contribution to this rise. But any rhetoric is tempered by Ettore's ironic character, of which we will see frequent examples in his letters. Regarding his nomination the previous day, he wrote the following to his mother:

> "Dear mother, . . . Today we purchased the furnishings for the office that the Faculty has graciously offered me. Practically, the Institute consists of just Carrelli, his old associate prof. Maione and his young assistant Cennamo. There is also a professor of geophysics who is

hard to discover. I found a letter from the Rector that had been laying around for a good 2 months announcing my nomination "for high fame of singular expertise". Unable to find him, I responded with a letter as high-minded as his..."

While earlier, on 21st November 1937, he expressed himself this way to his friend and colleague, the physicist Giovanni Gentile Jr.:

"Dear Gentile... I marvel that you doubt the strength of my stomach, in a metaphorical sense. Pope Pius XI is very old and I received an excellent Christian education. If at the next conclave they make me pope for exceptional merits, I will without doubt accept..."

Surprising, in any event, is his apparently sincere commitment to devote his energy to the Italian school and science, stated as it was just two months before March 1938.

When Ettore boarded the steamship, therefore, Ettore had in his pocket between 7,000 and 9,000 liras.[11] Liras of 1938, of course. But it is not the whole story: On 22nd January, again writing to his mother, Ettore requested: "Please

[11] Wages were paid by the Banca d'Italia (which was also open on Saturday morning) on the 27th, which that month fell on a Sunday. The salary for March, therefore, would have been already available on Saturday the 26th. But Ettore embarked on Friday, without waiting for it.

ask Luciano to withdraw my part from the bank account and maybe to send all of it to me, taking into account my previous withdrawals, and after returning to you the thousand lira that you recently loaned me". Therefore, when Ettore disappeared he possessed *at least* 10,000–15,000 liras, corresponding to at least 30,000 US dollars today.

The authorities closed the case reluctantly, as they were continuously urged to action, not only by Ettore's brothers, but by his cousin Francesco Maria Dominedò, the son of his paternal aunt Emilia. A year later, on 4th April 1939-XVII, the head of the Police's Border and Transportation Division, Saporiti, again asked if the report on Majorana should be maintained in evidence. The director of the Division of General and Confidential Affairs replied on the official form: "Action to be taken: *Disbar*".[12]

For us, on the contrary, the *Majorana Case* begins here.

---

[12] Document n. 8887- To the record on 22nd April 1939.

2

Fame

But then there are the geniuses like Galileo and Newton.
Well, Ettore Majorana was one of these.
ENRICO FERMI

Why Be Involved in Majorana?

Why concern ourselves with Ettore Majorana?

Majorana's fame is firmly justified through testimonies like the one below, which we owe to the mindful pen of Giuseppe Cocconi. Let us read it in full. From CERN in Geneva, Cocconi wrote to Edoardo Amaldi (a former collaborator of Enrico Fermi and colleague of Ettore):

> "Geneva, 1965, July 18th — Dear Amaldi, in a discussion that took place long ago on the book [*later published by the Accademia dei Lincei*] that you are writing about Ettore Majorana, I told you that I, too, had tenuous contact with Majorana just prior to the his disappearance. You expressed then that you wished me to describe my recollections in greater detail, so here I will try to satisfy you.
>
> After having just graduated in January 1938, I was offered, mainly by you, the opportunity to come to Rome for six months as an assistant at the university's

Institute of Physics. Once there, I was fortunate enough to join Fermi, Bernardini (who had taken a teaching post in Camerino a few months prior) and Ageno (himself also a young graduate), to engage in research regarding the products of the disintegration of "mu mesons" (then called mesotrons or yukons) produced by cosmic rays. The existence of "mu mesons" had been proposed about a year earlier, and the problem of their decay was quite fashionable.

It was indeed while I was with Fermi in the small workshop on the second floor, he working intently at the lathe a piece of a Wilson chamber that was intended to reveal the mesons at their end range, and I busy building a jalopy to illuminate the chamber using the flash produced by the explosion of a strip of aluminium on a shorted battery, that Ettore Majorana came in looking for Fermi. I was introduced to him and we exchanged a few words. A dark face, and that was all. It would have been a quite forgettable episode if, a few weeks later while with Fermi in the same workshop, I had not heard the news of Majorana's disappearance from Naples. I remember that Fermi busied himself by phoning various places until, few days later, one got the impression that no one would ever find him.

It was then that Fermi, trying to emphasize to me the significance of this loss, expressed himself in a rather unusual way for someone who was so severe when judging others. And at this point, I would like to repeat his words just as they echo in my memory: *"Because, you see, in the world there are various categories of scientists. People of second and third rank, who do their best but do not go very far. There are also people of the first rank, who make discoveries of great importance that are fundamental for the development of science* (and I have the distinct impression that he would have put himself in this category). *But then there are the geniuses like Galileo and Newton. Well, Ettore was one of these. Majorana had what no one else in the world has. But unfortunately, he lacked what is instead common in other men, plain good sense."*

I hope these lines provide you with what you wished to know. Kindest regards."

Giuseppe Cocconi.

"Plain good sense": we prefer to say *common sense*; which may not always be good, or the best.

Enrico Fermi, a 1938 Nobel laureate and one of the greatest physicists of our time (for his accomplishments in 1942 in Chicago his name will perhaps become as legendary as that of Prometheus…), expressed himself in an unusual way on another occasion, when he wrote from Rome on 27th July 1938, to Prime Minister Mussolini asking to intensify the search for Ettore: "I do not hesitate to declare, and this is not hyperbole, that of all the Italian and foreign scholars whom I had the opportunity to meet, Majorana is the one who for the depth of his genius has impressed me the most".

Bruno Pontecorvo, a direct observer, added: "A short time after his entry in Fermi's group, Majorana had already acquired so much knowledge and had reached such a level of understanding of physics, that he was able to speak with Fermi about scientific problems as an equal. Fermi himself considered Ettore the greatest theoretical physicist of our time. Often he was left astounded [...]. I remember Fermi's exact words: "If a problem has already been posed, no one in the world can solve it better than Majorana"."

Majorana, therefore, is famous not only for his disappearance. He was really a genius, and of a brilliance that placed him well ahead of this time. His fame, as happens with *true* fame, has grown over time, even among his colleagues. In fact, in recent decades it has exploded, and a large percentage of scientific papers from around the world (especially in the field of elementary particle physics) cite his name in their Title.

Enrico Fermi was perhaps one of the last examples, and an extraordinary one, of a great theorist who was also a great experimenter. Majorana, instead, was a pure theorist. Indeed (to use the same words as Fermi in the continuation of his letter to Mussolini), Ettore possessed, to the highest degree, that rare combination of skills that make up a theoretical physicist of *gran classe.* Ettore "carried" science, as Sciascia said; indeed he "carried" theoretical physics. He was certainly not less than a Wigner[1] or Weyl: who, for their aptitude in physics and

---

[1] Nobel Laureate, 1963.

mathematics, were perhaps the only scholars for whom Ettore himself harbored unreserved admiration.

So, on the one hand, he had no propensity for experimental activities (even if forced, to be clear, he could never have made a tangible contribution to projects like the technological construction of the atomic bomb). But, on the other hand, he could descend with unsurpassed and hardly imaginable depths into the substance of physical phenomena, seeing in them elegant symmetries and powerful new mathematical structures, or uncovering sophisticated physical laws. His sharpness enabled him to see beyond the vision of his colleagues: that is, to be a pioneer. Even his notes, written in 1927 when he began his transition from engineering studies to the study of physics, are not only a model of order (they are divided into topics and even have indexes), but also of originality, conciseness, and choice of the essentials. For this reason these notes, now known as the *Volumetti*, were suitable for publication just as they were. And indeed they have been published in 2003 by Kluwer Academic Press (in English), and in 2006 by Zanichelli (in Italian), thanks to the work of a few Italian and foreign scholars, besides our own. These "study" notes are in reality rich with original discoveries. This is not less true for the remaining manuscripts, which consist of scientific research notes only. But the publication of all these manuscripts is a heavy undertaking; and in 2009 we published with Springer (in English), in another 500 page volume, a *selection* of the so-called *Quaderni*: which contain many, but not all, of the scientific manuscripts left unpublished by Majorana.

Let us recall that Majorana, who switched to physics at the beginning of 1928, graduated[2] under Fermi on 6th July 1929; and obtained his "Libera Docenza" in theoretical physics on 12th November 1932.

From Galileo to Fermi

To provide an idea of what the activities of Fermi and his group meant to Italian culture and science, recall that the Italian physical sciences had once before achieved a position of international superiority — with Galileo. But while condemnation by the Church on 22nd July 1633, did not have serious consequences for Galileo himself, it proved disastrous for the Galileian School of physics, which could have continued on as the finest in the world. The vast, promising scientific movement founded by Galileo was cut at the roots by the condemnation of the master, such that science then transferred beyond the Alps. John Milton, recalling a visit to the "the famous

---

[2] Among the members of the Graduation Committee were Corbino, Fermi, Volterra, Levi-Civita, Lo Surdo, Armellini, and Trabacchi. The marks earned by Ettore during his studies were: Analytic and Projective Geometry, Rational Mechanics, Advanced Physics, Mathematical Physics, and Physics of the Earth: 30 out of 30 with honors; Algebraic Analysis, Descriptive Geometry, Experimental Physics, Exercises in physics: 30; Applied Chemistry: 28.5; General Chemistry, Infinitesimal Analysis: 27…

Galileo, by now old and a prisoner of the Inquisition" (Galileo died in 1642), summed up the situation brilliantly, noting in 1644 that *the state of servitude to which science had been reduced in its homeland was the reason why the Italian spirit — so alive before — was by now extinct, and for many years thereafter everything that was written was nothing more than flattery and platitudes.* Almost two centuries passed before another great physicist surfaced: Alessandro Volta. Volta created a branch of research that led to predominantly technological applications by Antonio Pacinotti, Galileo Ferraris, and Augusto Righi and, later, to those of Guglielmo Marconi. But this did not yield a true "school" of physics. So by the end of 1926, when Fermi obtained the chair of theoretical physics in Rome, Italy was certainly not prevalent in the world of physics.

It was only Fermi who, three centuries after Galileo, managed to generate an extensive, modern movement within the physical sciences. For example, the article which initiated Fermi's theory of "weak interactions" (coronated 50 years later, in 1983, by the experimental findings of Carlo Rubbia[3]) was released in 1933, exactly *300 years* after the final sentencing of Galilean theory. This digression into the past might help to clarify the cultural significance, as well as the difficulty, of the reconquest of Italian physics in the last century. In this context, the presence of Ettore Majorana was *potentially*

---

[3] Nobel Prize, 1984.

crucial. As we have demonstrated, he was beneath no other theoretical physicist or physico-mathematician. But, as a pioneer far ahead of his time (and also due to his shyness and self-critical nature), only a few of his articles were quickly understood, appreciated, and utilized. Moreover, science lost the man, his work and his leadership very early on. Certainly, if Italian theoretical physics had been able to draw upon the genius of Majorana (and his students), together with Fermi, for a longer period, the consequences could have been enormous, perhaps unimaginable. This does not change the fact that the fruits of Ettore Majorana's farsighted intellect can be seized — and are seized — *especially today* — not only in Italy, but all over *the world.*

The Rome Group

The previously mentioned rebirth of Italian physics would perhaps not have taken place at all without the involvement of Orso Mario Corbino. A Sicilian, he graduated in physics in Palermo and was first a university professor in Messina, then in Rome. In 1918, he became the director of the Institute of Physics in Via Panisperna street in Rome. At the suggestion of Prime Minister Giolitti, Corbino was appointed senator in 1920, and the following year became the Minister of Education. In 1923, Mussolini made him Minister of the National Economy in the first government over which he presided. When Corbino met Fermi, he recognized that he could finally realize his dream of taking a quantum leap in Italian scientific research.

"I met Senator Corbino," said Fermi in 1937, "when I returned to Rome after having just graduated in 1922. I was then twenty years old, Corbino forty-six; ... and he was universally recognized as one of the most eminent personalities in this field of study at that time... We had almost daily conversations and discussions in those days, and as a result... I developed the deep and heartfelt devotion of a disciple to his maestro."

In 1926, while Fermi was in Florence (actually, in Arcetri; the florentine laboratory was located near the villa where Galileo spent the last years of his life) to carry out mainly experimental research with his friend and contemporary Franco Rasetti — a man gifted with exceptional ability and versatility as an experimenter —, Corbino conceived a professorship in theoretical physics at the University of Rome that was tailor-made for the 25 year old Fermi. Corbino protected him from the inevitable envy of colleagues who would otherwise have given him very little space. He endeavored to recruit and place his group members academically, until in 1929 he proposed Fermi as the official leader of physics of the period by getting him elected to the Royal Academy of Italy. That same year, Corbino and Fermi took the historic decision to concentrate the group's efforts on the nascent field of nuclear physics. Corbino made the announcement in a major speech, highly prophetic, to the Italian Society for the Advancement of Science; and afterward guaranteed the Rome group financial support which was exceptional for that period.

Meanwhile, at the end of 1923, the 17 year old Majorana enrolled at the University of Rome to study engineering (he had lived in Rome since the age of 8 or 9, along with his brothers and cousins, at the Jesuit boarding school "Convitto Massimo alle Terme" until 1921, when the entire family moved to Rome and Ettore went on from being a boarder to an external student). His schoolmates included his brother Luciano, Emilio Segré, Gastone Piqué, Enrico Volterra, Giovanni Gentile Jr, and Giovanni Enriques. The latter three were, respectively, sons of the great mathematician Vito Volterra (one of the few university professors to refuse the oath of allegiance to the fascist regime), the philosopher Giovanni Gentile (whom we have already mentioned), and the well-known mathematician and epistemologist Federigo Enriques. In June 1927, Corbino launched an appeal to his engineering students in the hopes that, with the appointment of Fermi in Rome, the most talented among them would switch to the study of physics. Edoardo Amaldi, then at the end of his second year, accepted this invitation. Almost simultaneously, Segré met Rasetti and then Fermi, and he too decided to switch to physics. Here, Segré began to speak of the extraordinary gifts of Ettore Majorana, and one day convinced him to meet with Fermi. Ettore's transfer to physics took place at the beginning of 1928 — that is, at the beginning of his fifth year of university — following a conversation with Fermi narrated in detail by Segré, which we will discuss later.

Amaldi recounts: *It was then that I saw Majorana for the first time. From a distance he looked slim, with a shy and uncertain gait. Up close you could see his jet black hair, dark complexion, slightly hollowed cheeks, sparkling and vivacious eyes: the overall appearance of a Saracen.* Amaldi was the first historian of Ettore Majorana, and in the Appendix we include what he has written on the subject (according to the version that appeared in 1968 in the *Giornale di Fisica*, which has been simplified by us in the sections related to scientific subjects).

Later, Bruno Pontecorvo joined the group, which now consisted of E. Fermi, F. Rasetti, E. Majorana, E. Amaldi, E. Segré, and B. Pontecorvo, in addition to chemist O. D'Agostino. It is well known that at the Institute in Via Panisperna, Corbino was nicknamed "The Almighty", Fermi "The Pope", Rasetti "the Venerable Master" or "Cardinal Vicar", Segré "Basilisk" (for his biting character), and Majorana "The Grand Inquisitor", for his critical personality, of course.

Fermi would receive the Nobel Prize in 1938.

Of the Rome group, only Amaldi remained *in patria*, taking upon himself the major burden of keeping alive in Italy the school of physics created by Fermi; and later on he was one of the fathers of the

European laboratories in Geneva (CERN). The others emigrated: Fermi and Segré to the United States (where Segré would receive the Nobel Prize in 1959 for his contribution to the discovery of the antiproton), and Pontecorvo in 1950, after a long period in the US, Canada, and England, to the Soviet Union (where, universally known as *Bruno Maksimovich*, he soon became an Academician of Sciences of the USSR, and one of the directors of the leading laboratories in Dubna).[4] Rasetti, however, after having achieved such great success as the principal experimentalist of the Rome group, later abandoned physics completely, becoming instead an internationally renowned paleontologist, and later an acclaimed botanist. His personality was so extraordinary that we would like to include here a letter written from Waremme (Belgium) on 22nd June 1984, penned when he was nearly 80 years old, in response to a standard request for "testimony":

> "Dear Professor, — I beg your pardon for replying only now to your letter of April 4th of last year. It arrived while I was away on a trip for botanical purposes from May 6th to June 17th; so I have looked it over only a few days ago. At the moment I do not have the time to make all of the comments the subject deserves, so I'll

---

[4] Jean-Pierre Vigier argues that Pontecorvo emigrated (in September 1950) to Russia to help balance the global scientific positions: a "political" decision, therefore, as well as scientific.

relate to you only my first impressions while resolving to write to you a more detailed account later.

I could not attend the "Fermian"[5] conference because I was ill for a month and a half, which caused me to abandon a trip to Palermo for which I had already purchased an air ticket; a trip for the purpose of photographing the "Orchidaceae" characteristic of the province of Palermo. I made another trip, the one stated above, to the south-central part of Italy. As you know, many years ago I lost all interest in physics, devoting myself with great success and satisfaction first to Cambrian-period paleontology of Canada, United States and Sardinia, then to the botany of the Alps, and finally to Italian Orchidaceae.

Even from a first, brief overview of your letter, I understand that you have information and knowledge of Ettore's life which is far more comprehensive and thorough than what has been written by others, who possessed much more imagination than seriousness.

...I'd like to point out an incredible "gaffe" made by the Corriere della Sera [*of 13 Dec 83, page. 16*] in the caption under the famous photograph called "the three priests" (because a friend of Laura Fermi, seeing a copy hanging on her sitting room wall, exclaimed: "Who are those three priests?").[6] From the caption

[5] The conference was held on 26th April 1984 in Bologna on the occasion of the 50th anniversary of the formulation of Fermi's theory on "beta decay" (*Cinquant'anni di fisica delle interazioni deboli*), which will be discussed in what follows.

[6] The photograph in question is reproduced in this book.

of the Corriere, it would seem that the photograph represents Fermi, Rasetti and Segré as senior grad students! Rather, Fermi and I graduated in 1922, when the Fascist regime did not yet exist and undergraduates did not wear black shirts! The photograph was instead taken in the mid-thirties, and all three of us were members of a graduation committee, not students ourselves... If such monstrosities are present in what should be considered a quite serious newspaper, what should we believe of what is written in less reputable journals?

With this I conclude my comments, for the moment. Please accept my best and kindest regards.

Franco Rasetti

PS: I neglected to mention that it would be vain to expect from me any information regarding Majorana that is not already well known. Of the physicists of Via Panisperna, I am certainly the one who knew him least. In any case, I was somewhat older than him, while Segrè and Amaldi were more or less the same age. Furthermore, I was connected to these through sporting activities, while Majorana was totally foreign to any sport. I used to play tennis with Segrè and Amaldi, but mostly we were mountain climbing companions (for example, we completed several difficult climbs, always without a guide, including the crossing of the Matterhorn)."

We cannot leave the "Rome group" without at least mentioning that of Florence, where physics was flourishing at around the same period. The same competition, which gave Fermi the Rome professorship, at the end of 1926 conferred one also to Enrico Persico in Florence, and Aldo Pontremoli in Milan.[7] Enrico Persico had a strong and beneficial influence. In the words of Segré, "In Florence Persico was the only tenured professor of the new generation, and taught with great success. The young physicists who were in Florence and took on modern problems were: Bruno Rossi, famous for his study of cosmic rays, Giuseppe Occhialini, who worked resolutely on the discovery of both clusters[8] and the pion; Gilberto Bernardini, later the scientific director of CERN in Geneva and director of the Scuola Normale Superiore of Pisa, and Giulio Racah, inventor of the coefficients that bear his name and who later became President of the University of Jerusalem. Although these Florentine physicists were in their early twenties, they were full of enthusiasm, energy and ideas. The groups from Florence and Rome were bound by a

---

[7] Pontremoli later perished in the crash of the airship *Italia* during the "Noble expedition" to the North Pole.

[8] Segré refers to the fact that Occhialini actively collaborated (along with the Englishman P. Blackett) in the discovery of the first anti-particle, the positron, and the phenomena of creation and annihilation of electron–positron pairs. But the Nobel Prize was awarded (in 1936) only to the American, C. D. Anderson who had revealed the positron slightly before. With Blackett it was remedied by giving him the Nobel Prize on another occasion (in 1948). As to the discovery of the first meson, the *pion* (the particle theorized in 1934 by H. Yukawa), this was made in 1947 by C. Lattes, G. Occhialini and C. Powell, but once again the Nobel Prize was given only to the Englishman Powell (in 1950). With the Italian Occhialini (and the Brazilian of Italian origin Cesare Lattes), however, no "remedy" from the Swedish Academy of Sciences was ever forthcoming.

genuine friendship, and often exchanged visits and seminars."

Competition for the 1937 Professorships

After the competition of 1926 in which Fermi, Persico, and Pontremoli received their professorships, another 10 years passed before a new competition opened, in 1937, for theoretical physics at the University of Palermo at the request of Emilio Segré.

The events of this competition, and especially its antecedents, gave rise to a heated argument in 1975 between Leonardo Sciascia, Edoardo Amaldi, and others (including Segré and the author). Here, we will restrict ourselves, following our inclination, to reproduce the certain documents in our possession — this time, courtesy of the brothers Dubini, from Ticino Canton (Switzerland), who reside in Cologne, Germany. The competitors were numerous and many of them highly qualified; four in particular: Ettore Majorana, Giulio Racah (who was Jewish, and later moved from Florence to Israel, where he founded the Israeli theoretical physics school), Giancarlo Wick (whose mother was a known anti-fascist in Turin), and Giovannino Gentile (the son of the philosopher with the same name, the former Minister of Public Instruction as we would say today) and who later became the inventor of "parastatistics" in quantum mechanics. The distinguished Judging Committee consisted of Enrico Fermi (chairman), Antonio

Carrelli, Orazio Lazzarino, Enrico Persico, and Giovanni Polvani.

Record no. 1 reads as follows:

> "The Committee appointed by S.E.[9] the Minister of National Education, and consisting of Professors Carrelli Antonio, Fermi S.E. Enrico, Lazzarino Orazio, Persico Enrico, and Polvani Giovanni, assembled at 4 o'clock p.m. on October 25th, 1937 in a lecture hall of the Institute of Physics at the University of Rome. The Commission was established with S.E. Fermi appointed as President, and Carrelli as Secretary.
>
> After a thorough exchange of ideas, the Committee is unanimous in recognizing the highly exceptional scientific standing of Professor Ettore Majorana, who is one of the candidates. Therefore, the Commission decides to send a letter and a report to S.E. the Minister to outline the opportunity of appointing Majorana a professor of Theoretical Physics for high and deserved reputation within a University of the Kingdom, independent of the competition required by the University of Palermo. The Commission, awaiting instructions from S.E. the Minister, is adjourned until a new convocation.
>
> The assembly concludes at 7 o'clock p.m. Read, approved and signed on the spot."
>
> E. Fermi, O. Lazzarino,
> E. Persico, G. Polvani, A. Carrelli

[9] Abbreviation for "Sua Eccellenza": *His Excellency* in English.

The letter, sent to the minister the same day, and on which Minister Bottai hand-wrote the word "Urgent", repeats the contents of the previous record, declaring Professor Ettore Majorana to have, compared to the other candidates, a national and international scientific standing such that "the Commission hesitates to apply to him the normal procedure for university competitions." This letter has an attachment, *Report on the scientific activity of Prof. Ettore Majorana*, signed, as always, in the following order: Fermi, Lazzarino, Persico, Polvani, and Carrelli. Let's have a look:

> "Prof. Ettore Majorana graduated in Physics in Rome in 1929. Since the beginning of his scientific career, he has demonstrated a depth of thought and conceptual genius that has attracted the attention of scholars of theoretical physics around the world. Without listing his works, which are all quite remarkable for the originality of the methods employed and the importance of the results achieved, we limit ourselves here to the following notices:
>
> In modern nuclear theory the contribution made by this researcher with the introduction of forces known as "Majorana Forces" is universally recognized as being, among the most important, the one which allows understanding theoretically the reasons for the stability of nuclei. Majorana's work is today the basis for the most important research in this field.
>
> In atomic theory, Majorana deserves the credit for having solved, with simple and elegant symmetry

considerations, some of the most intricate questions regarding the structure of the spectra.

In a recent paper he devised a brilliant method for treating positive and negative electrons in a symmetric way, finally eliminating the need for the highly artificial and unsatisfactory hypothesis of an infinitely large electric charge diffused throughout space, a question which had been unsuccessfully confronted by numerous other scholars."

One of Ettore's most important works, the one in which he introduces his "infinite-components equation" (we will discuss this in Chapter 5), is not mentioned: It had not yet been understood. It is interesting to note, however, that due attention is given to his symmetrical theory for the electron and the anti-electron (today so much in fashion for its application to neutrino and anti-neutrino, as well as for the introduction, e.g., of the Majorana Fermions recently discovered in important experiments in Condensate Matter Physics, and recognized to be essential tools even for Quantum Computing) because of its ability to eliminate the assumptions of the so-called "Dirac sea"[10], hypothesis, which is defined as "highly artificial and unsatisfactory" despite having been, by most people, accepted uncritically. That touch of originality in a bureaucratic document cheers up, and we find ourselves in agreement with that argument.

---

[10] P.A.M. Dirac, Nobel Prize, 1933.

Once the professorship was conferred to Ettore "out of competition",[11] the Committee resumed its work, reaching consensus in the formation of a triple winner: (1) GianCarlo Wick, (2) Giulio Racah, (3) Giovannino Gentile.

Wick went to Palermo, Racah to Pisa, and Gentile Jr. to Milan.

Giovannino Gentile, a close friend of Ettore, would die prematurely in 1942.

---

[11] By applying a law that had already been used, for example, to award a university professorship, out of competition, to Guglielmo Marconi (Nobel Prize, 1909).

3

The Family

The Majoranas are like potatoes: the best part is buried.

The Grandfather: "Founder" of the Family

"The Majoranas are like potatoes: the best part is buried," was occasionally heard in the Majorana household. But the standard set by Ettore Majorana's ancestors — ministers, deputies, rectors, orators, scientists — was actually quite a heavy burden to bear for the Majorana grandchildren and great-grandchildren…

A family legend, to which no one gives real credit, goes back as far as Julius Valerius Maiorianus, proclaimed emperor of the western Roman empire in Ravenna in the year 457 A.D., who was murdered by Ricimer in 461. Today, the name translates to "Maggioriano", but appears as *Majorano* in less recent publications.

With greater realism, let us return to a time nearer our own, with the help of the *Dictionary of Famous Men* published in Italian in Naples in the year 1761, and bearing the complete title of *Historical Dictionary, portable, containing the story of Famous Men in the arts and Sciences, with their major Works, and the best Editions of them. Written in French by Mr. Abate LADVOCAT. Very new Edition…*

with a supplement by GIANGIUSEPPE ORIGLIA PAULINO inserted in its places, and with notes by Father ANTONMARIA LUGO, Somaschan, now corrected and improved too. Tome IV. Dedicated to Her Excellency Lady AUGUSTA CATERINA PICCOLOMINI of the Free Lords of Triana, Sienese Patrizia, Duchess of Vastogirardi, Marchise of Caccavone, helpful Lady of the manor of Panicocoli, of Macchia Bovina, Saint Mauro, Saint Lenci, Spring of Paradise, High Peaks, Pizzuoli, Cocozza, Cocozzella, St. Elizabeth, Civitella, Quarticciolo, Cerrito, Bralli and Luciso & c. In Naples, MDCCLXI. At Benedetto Gessari. Under license — of course — *of Superiors, and Privilege.* Volume IV of this Dictionary records in ascending order of interest (for us): "MAJORANA (Fulvio). Neapoletan Patrician, Juriconsult of the XVII century. Gave birth to: *Opopraxis criminalis: De Poenis etc.*"; MAJORANA (Piero) Jureconsult, from Palermo, died in 1709. Wrote: *Selecta Hypothecaria et Feudalis etc. De Jure Taren Possessionis Tractat*; and finally "Salvadore MAJORANA, from Palermo as well, was a Poet, who flourished around 1600, and left: *Canzoni Siciliane*". The latter two are Sicilian; the name of the last one is Salvatore, as for the known "founder" of the family. And, on the back of the title page, there is a small surprise; the *ex-libris* (book-plate) says: "Jacobi M[a]. Canonici Magro ex Baronibus M. F. Nichiarae etc."; *della Nicchiara* is the title of one of the branches of the Majorana family.

The true forefather of Ettore's family was, in any case, his grandfather Salvatore Majorana Calatabiano (son, precisely, of Valentino Majorana and Antonia Calatabiano), born in Militello Val di Catania on Christmas Eve, 1825. "He was a prime example of a self-made man," said Vittorio Emanuele Orlando about Salvatore at the end of 1897 in commemorating his death. "Born from virtually nothing," others wrote, "he was an economist, a lawyer, a thinker, a writer, a teacher, parliamentarian and statesman, a government leader, and a patriot...; on top of that, the speech: which for him was a form of austere and complete eloquence, all made out of ideas, irresistible." At 21 he publishes his first book, *Wealth and Poverty*, a treatise on political economics which, in supporting economic and social freedoms at the beginning of 1848, was hailed as a scientifico-economic justification for the revolution. Unpopular for that with the Bourbons of the Restoration, nevertheless in 1857 he would successfully defend conspirator Luigi Pellegrino in front of the High Court of Justice. In 1865, he is appointed full professor at the University of Messina (and shortly thereafter of Catania), and the following year he is elected Member of the national Parliament for the district of Nicosia; here, becoming one of the leaders of the Left despite the fact that old traditions, personal interests, and ancient deeply-rooted factions in Militello tenaciously oppose "the young man who, representative of new energies, appeared in that village with a radically liberal program". In 1869, in fact, they go so far as to murder his stepson Francesco, whose mother (the widow Laganà) had been Salvatore's first wife. As instigator of the murder, the Baron Majorana Cocuzzella, a Member of Parliament for the constituency of Militello, is arrested and then acquitted — obviously (we cannot refrain from saying) — by the Court d'Assises of Catania.

In the Chamber of Deputies, he devotes his energy to the reorganization of credit as well as other economic and financial issues, gathering for a project of his the support of the whole of the Left, from Crispi to Cairoli. He opposed the tax on flour, and, following the 20th September 1870 military expedition which led to the conquest of Rome and the unification of Italy (for which he was a promoter), he begins a arm wrestling with Prime Minister Sella, which will end only with the fall of the Right and the election of Depretis as Prime Minister.[1] "When Majorana speaks," wrote Milan's newspaper *Il Secolo*, "there is no rapporteur who can keep up; he is the despair and at the same time the wonder of the stenographers, who end up throwing away their pencils and doing as the others did, that is, *watching* the orator… While the oratory of Majorana Calatabiano can make one giddy, reading his words reveals what an educated man and a scientist he was." This was followed by the *Gazzetta di Torino* in 1875: "When he

---

[1] In the meantime, through his publications, he shares with Francesco Ferrara the task of transplanting and fortifying in Italy the "Liberalism" doctrines, turning economics into a political science.

speaks, motion ceases, chatter stops. He has a sonorous voice, his speech is agile. No one can raise questions up there, more than he... Majorana is always masterful, as he remains ever within the dazzling atmosphere of science and principles." What follows may interest us; he was convinced that economic laws were natural laws with a mathematical basis, maintaining that

> "it is a disregard of the dictates of science — that in the end should be in the legislative domain what in the technical applications are the theorems of physics and calculus —, it is this very disdain, this divorce between thought and practice, between science and social art, the prevailing cause of the distress in which the State finds itself."

In the first ministry of the Left, Prime Minister Depretis entrusts Salvatore with the portfolio of Agriculture, Industry, and Commerce. He receives from the King, Vittorio Emanuele II, the Grande Cordone of the Crown of Italy, and the Legion of Honor from France. After the crisis of the end of 1877, he returns to his post as minister of Depretis' third government, and begins a project to reform the currency issuance institutions which keeps him very busy, clearing the way for a true "income-based economy" but being prejudicial to the interests of the privileged banks (the Banca Nazionale, in particular). He obtains approval for his banking law, but at such a price that he considers returning solely to university teaching and family life. Nevertheless, in 1879 he accepts an appointment to the Senate, to which he devotes almost 20 years, until his death.

Members of Parliament, University Rectors, Scientists: His Paternal Uncles

Following his marriage to his second wife, Rosa Campisi, Salvatore has seven children: Giuseppe, Angelo, Quirino, Dante, Fabio Massimo (Ettore's father), Elvira, and Emilia. Children who, with the help of his wife, he educates all "to the unswerving habit of hard work." In the periods spent with the family, *rising very early in the morning*, the children will remember, *he would be at home studying — himself an example —, and wanted his children to study with him at his table, or one nearby.* This would continue well into the evening, *such that it became a habit for him to dine with his children in the evening at a late hour.* The methods employed by these parents, perhaps because they were accompanied by dedication and affection, are apparently successful. Three of the children become parliament members and Rectors of the University of Catania (yes, three brothers became rectors of the same university, respectively in the years 1895–1898, 1911–1919, 1944–1947). If forced to choose among them, we should focus particular attention on Angelo, who would first become Minister of Finance, then of the Treasury. We will do this later.

Quirino graduates at age 19 in Engineering and at 21 in Physical Sciences and Mathematics, and later becomes president of the Italian Physical Society.

This time we will allow the Treccani Encyclopaedic Dictionary (which, incidentally, includes several Majorana: *all* of whom are descendants of Salvatore) to speak for us: About Quirino, who was still alive at the time, it said: "Physicist (born: Catania 1871), brother of Angelo; director of the Higher institute of telephones and telegraphs of the state (1904–1914), then professor of experimental physics at the Polytechnic of Turin, and (since 1921) in Bologna where he succeeded A. Righi as director of the physics institute. He is a member of the Accademia dei Lincei… He has conducted important research in experimental physics (on cathode rays, the Volta effect, photoelectric phenomena, the constancy of the speed of light emitted by a source in motion, etc.). He has achieved notable results in the field of telecommunications, performing numerous experiments in wireless telegraphy over great distances, as well as optical telephony with ordinary, ultraviolet and infrared light. In the course of these experiences, he conceived a microphone… (*the hydraulic microphone of M.*). He was also busily engaged in a number of issues related to the theory of relativity, and in particular in the possible absorption of *gravity* in matter." And we will add that he was also a member of the commission that conferred the professorship to Enrico Fermi; he discovered magnetic birefringence; wrote his first book (on X-rays) at 25 years of age (Rome, 1897); performed some of the most refined and original experiments on gravity known to us, and he devoted considerable energy to demonstrating the falsity of the theory of Special Relativity: but, as a skilled and rigorous experimenter, he would only *confirm* Einstein's theory. He dies in 1957. Overall, a great scientist indeed, but not comparable, of course, to Ettore.

Elvira and Emilia complete their education in Rome, and there they marry (the first to State councilor Savini-Nicci, the second to the attorney Dominedò).

His Father

Even Fabio, Ettore's father (b. Catania, 1875–d. Rome, 1934), graduated very young — at 19 — in Engineering, then in Physical and Mathematical sciences. He would take it upon himself to provide the young Ettore (who completed his first years of elementary school at home) both his cultural and scholastic instruction until he reaches the age of eight or nine years, when, as mentioned previously, Ettore is sent to the officially recognized Jesuit boarding school in Rome to finish elementary school and attend middle and high school. Ettore would always remain very attached to his father, and will be undoubtedly affected deeply by his death in 1934.

Ettore's cousins, Angelo and Salvatore (sons of Dante), recalled for instance that at eight years of age Ettore already displayed well above average abilities. It is rather known that at seven he was a skilled chess player (as the local newpaper noted); while at four, while hiding under a table out of shyness, he would calculate in mere seconds the products of two three-digit numbers and square or cubic roots...as a game. Till recent times one could still see the numbers that Ettore scribbled in pencil on the walls of the balconies of the Majorana house in Via Etnea in Catania, written in a child's uncertain scrawl, when he was just a few years old.

Among the few documents remaining from his boarding school period, we have a Christmas card (stamped Rome, 24.XII.1915) to Ing. Fabio Majorana, Catania: *Dear parents: Greetings to all and merry Christmas. You will receive a kiss from — your loving son, Ettore.* And two undated letters, although written around the same period: *Dear father: I received the letter from mother. When will you come? We* [also referring to his brothers and cousins] *are all well. I have not received your letters. How are you and mother? Has grandmother returned? Will Maria also come? It rains here quite often. A million kisses — Your loving son, Ettore.* And finally: *Dear mother — We are all well. When will father come? Will you come for Carnivale? Often we cannot go for walks due to bad weather. Please also bring grandmother to Rome. Did you get my letter? Valentino* [Dominedò] *has recovered and has been well for many days.* Perhaps he was writing just during the bad weather; and the statement that he could not "go for walks" appears to be a veiled attempt to elicit sympathy from his mother. Soon after, however, he addresses the many questions from his mother courteously, but with disguised sufficiency; as for problems somewhat tiresome and almost business-like: *It seems to me that in the letter I sent you I wrote all that you asked me. If you want to know anything else, write me immediately. A million kisses — Your loving son, Ettore.* (And earlier, when he writes "*you* will receive", was he referring only to his father, with *you* actually intended as singular?; so it may appear from the original card, in Italian...).

The Engineer Fabio Majorana founds the first telephone business in Catania and as a result, in that city his name becomes synonymous with "telephone company". Moving to Rome, in 1928 he is appointed as a division head for the Ministry of Communication, and a few years later becomes its inspector general. He also dedicates himself to structural engineering, and a text on the *Liberty* architectural style in Catania includes a picture of the family home he built in Via Sei Aprile.

But let us return for a moment to Ettore's uncles, the university rectors and parliament members. The eldest, Giuseppe (b. Catania, 1863), a highly cultured man, from the scientific standpoint engages in economic statistics, and publishes numerous books through the Loescher house in Rome, including: *Theory of the Value* (1887), *Statistics and the State Economy, The Large Number Law and Insurance, Theory of Statistics* (all in 1889),

The Natural Laws of Political Economy (1890), *Principle of Population* (1891) *and Statistical Data in the Banking Problem* (1894). The latter two in particular were highly praised in Germany, France, Croatia, and Spain; and the last one was digitized by Columbia University almost 120 years later (in 2009). He enjoys additional success abroad with a book first published in Florence (Barbera pub., 1889) and then in a more complete form in Catania, under the title *Programs of Theoretical and Applied Statistics* (1893). Also the author of several works of literature, later on he repudiates them, endeavoring to destroy all copies that he could find.

The youngest of the three, Dante, an attorney, devotes himself primarily to Law. We recall, for example, the books of his still typeset by E. Loescher & C., of Rome: *Hunting and its Legislation* (1898), *The Commune Hamlets in Italian Administrative Law* (1899), The *Legal Concept of the Science of the State* (1899), etc. Dante was elected to parliament in 1924.

But, of these three uncles, the most perplexing is without doubt Angelo, the lawyer and sociologist. He shines very early, but his light soon dies out. And the parable of his life leaves us thoughtful when we compare it to what we know of Ettore's.

His Uncle: The Minister

Angelo is born in Catania, the second of the seven sons of Salvatore, in December 1865. Under his parents' austere, demanding but loving guidance, he soon proves to be even more precocious than his brothers. He receives his high school diploma at age 12, and at 16 obtains his university Degree in Law in Rome, where he works as secretary to his father and where he became fondly regarded by the King and Queen of Italy. A patriot, he is refined, affable, and willing, and develops friendships with D'Annunzio, Minguzzi, and Oberdan. Between the ages of 18 and 20, he sends to the press his first works: *On Parliamentarism: Evils, Causes, Remedies* (Rome, 1885); *The Principle of Sovereignty in the Constitution of States* (Rome, 1886); *Constitutional Theory of Revenues and Expenditures of the State* (Rome, 1886). But, having acquired the qualifications to teach at the university level ("Libera Docenza") at age 17, he becomes already an "equalized professor" at the University of Catania. In 1886, he participates in three Competitions for professorships in Constitutional Law at the universities of Catania (with Silvio Spaventa the president of the selection committee), of Messina, and of Pavia, winning all three. He astoundes the selection committees, who acknowledge that "the winner, Mr. Angelo Majorana, lawyer at the Court of Cassation of Rome, has not yet reached his majority", that was — in those times — the age of 21. He then becomes a full professor in Catania, where he would be elected the University Rector at 29. He publishes several other books, about *The Historical*

Evolution of the Relations between Legislation and Jurisdiction (Bologna, 1889), and *The System of the Legal State* (Roma, 1889); writing thereafter principally on sociology: *The First Principles of Sociology* (Rome, 1891) and *Sociological Theory of the Political Establishment.* This latter work in particular, which emphasizes the ancient teachings of Vico and Romagnosi, gets translated into several languages and earns him an appointment as a member of the International Institute of Sociology, in Paris.

Hearty, handsome, and a seductive speaker, he influenced or charmed those around him; naturally inspiring confidence and solidarity, if not admiration. Indeed, he possessed the rare ability to make others forgive him for his talents. For his part, he was attracted to good-looking women (the family still recounts how, while a Minister and surrounded by colleagues in his office, upon catching sight in the post of a letter from a Roman noblewoman who finally conceded to him a long-awaited appointment, he leapt onto the table shouting "hooray"!).

At 28, he enters active politics, immediately making a name for himself: as evidenced by a parchment from Palazzo Marino which conveys praise from "The Municipality and the city of Milan". With a liberal slant ("the state must be the guarantor of all freedoms, and of the freedom of all"), from 1896 he gives himself up to politics. Prime Minister Giolitti entrusts him first with the position of Undersecretary for Finance, and then at 38 (in 1904), with the Ministry of Finance itself. Two years later, he is again a minister under Giolitti, this time of the Treasury, a portfolio around which revolved much of Italian politics at the time. He commits himself fully to this appointment, contributing to the revival of Italian finance, working on tax reform, and advocating an equitable distribution of public works throughout the nation. But not long thereafter, he begins his decline. Exhausted from excessive activity, it is with the same swiftness, that his genius had manifested, that his body yields to illness (nephritis). In May 1907, he leaves the government for a period of repose in Sicily. But he does not return to Rome; after three years, he passes away in Catania at only 44 years of age; but not before writing his final book (*The Art of Public Speaking*: chosen by Luigi Capuana as the topic of his inaugural lecture at the university that year, 1910), in which he is able to temper the idealistic impulses of his upbringing with the positivism that his studies had imparted in him.

We have lingered on the precocious, brilliant rise and swift decline of "Uncle Angelo"; and we had good grounds for it. But we cannot resist the temptation to provide a portion, an example, of one of the many speeches given by Angelo Majorana to the Chamber of Deputies: on the problems of the University. The title does not mince words: *The question of the "misfits", and the reform of public education.* We will provide the date later.

"Allow me the House to refer to some of the figures calculated by colleague Ferraris in his valuable statistical analysis; and that it is I who does this, being one who comes from the very southern provinces for which these figures give rise to less pleasurable considerations... We have already seen that we pour from our universities each year, into the public marketplace of national life (if you will pardon the economic-style phrasing), a good 1,070 law school graduates: that is, more than double what is needed! Similarly, while the requirements for physicians and surgeons is just 500 per year, the graduates in medicine and surgery, on the basis of an average which is always growing, are 928 per year!...

"Well then, the number of students enrolled in Universities and Upper institutions varies among the different regions and various Faculties in the following manner. In northern Italy, we have in the Law faculty a number of enrollees estimated at 13.85 for every 100,000 inhabitants; in the southern mainland, a share of 24.14, and in Sicily of 22.82. In the Medicine faculty, then, for every 100,000 inhabitants, we have a share of 17.73 in Upper Italy, compared to 19.06 in Sicily, and a good 27.26 in the Continental South. *(Murmuring)*...

"I add that, observing the enrollees in the Science faculties, and therefore also in the Schools of Engineering, an inverse phenomenon is revealed, though identical in substance. In the northern provinces, for every 100,000 inhabitants we have a share of 9.08, while in the south, the number is 5.39, and in Sicily it is 7.72...

"We ask ourselves: But all of these graduates above what is needed, all these lawyers who have no clients, all these doctors who have no patients, all these teachers who cannot find schools or salaries, what will they do within the very society of which they are a part? And do I have to say it? But who among us, questioning himself, cannot provide an answer? Nor do I need to tell you, honorable colleagues, that if there are no doctors who invent diseases to be able to cure them, unfortunately there are lawyers, especially in civil matters, inventing lawsuits to be able to defend them *(Laughter)*... Nor do I need to repeat how, with ill-fated consequences, the number of candidates competing for public office has multiplied; and how, precisely for this reason, the State and the Provinces and Municipalities and all public Administrations are forced to increase their competencies and offices and duties, together with the corresponding employees, greatly aggravating their budgets and thereby injuring, indirectly, the budget of the nation...

"This evil is not unique to Italy, but is also found in other regions. In Germany, famous are the adages of Virchow, who called the Universities the "seedbeds of misfits," and Emperor William, who defined the students of classical schools as "candidates for starvation."

And later, speaking against the centralistic bureaucratization of higher education, he concluded: "I find it utterly wrong to submit young people who are at the peak of their strength, who are usually above twenty years of age, that is, at the age of full responsibility and legal capacity regarding both civil and political rights: I find wrong, I say, that

> we would still submit them to such a *constrained* system of examinations, one which tends to suppress their sense of initiative, without exciting the challenge of healthy and fruitful study... Previously, I reminded you that most of our university student population sees a mirage: public office; has an ideal: payday.[2] Now then, note now what single psychological comparison runs between the payday, which is the future ideal, and passing the examination,[2] which is the nearest ideal! I frankly declare that I am against annual exams, while I am in favor of a final "state exam"... It is necessary to remove this humiliating tutelage, useless for the good, effective for the encouragement of indolence, which is placed on our youths by the current regime of university centralization..."

Without going into the merits of the ideas he expressed, it is difficult to believe that this speech was delivered much more than a century ago, during the session of 11th March 1899.

His Mother and Brothers

Edoardo Amaldi writes: "The marriage of Eng. Fabio to Mrs. Dorina Corso (b. Catania, 1876, d. Rome, 1966), also from Catania, produced five children: Rosina, who later married Werner Schultze; Salvatore, a doctor of law and scholar of philosophy; Luciano, a civil engineer

---

[2] The 27th was the day of the month that employees received their wages. On the other hand, 27/30 was and is a good grade for passing a university exam.

specialized in aeronautical construction but who then devoted himself to the design and construction of instruments for optical astronomy; Ettore (born 5th August 1906 in Catania); and fifth and last, Maria, musician and piano teacher."

The given name of Mrs. Corso was actually *Salvatrice*: a name not particularly pleasing, so Ettore's mother skilfully transformed it (inspired by the Spanish, and Portuguese, version of the name) into *Dorina.*

Salvatore left behind an enormous quantity of manuscripts, each filled with his philosophical and religious reflections.

Luciano designed (among other things) the Observatories of Monte Mario (Rome), of Gran Sasso (L'Aquila), and of Mount Etna (Catania). He is also known to have drafted a project to build a bridge across the Strait of Messina. He married Mrs. Nunni Cirino and had three children: Fabio Jr., Ettore Jr., and Pietro, who are currently the closest relatives of Ettore Majorana (nephews of the first degree, along with Fabio Wolfgang Schultze).

Maria, who graduated with full marks in piano from the Conservatory of Santa Cecilia in Rome and studied singing, was romantic and sensitive, but at the same time concretely and deeply humane. She resided in Rome. Enthusiastic for refinements and cultivated things, she devoted many of her years to fine arts; always leaning on a solid cultural background: never strayed far, that is, from the common sentiment of life. At her house in Rome, it was commonplace to hear her discuss — with keen safe judgment — prose, poetry, theater, and painting; and to listen to her playing Schumann and Debussy on the piano, or recite with delicate artistry the poetry of Mallarmé, Verlaine, Prévert. She performed some theater (Chekhov, for example) along with, and for the entertainment of, groups of impassionate friends. In one of her public performances, rare to the point of being unique, she remembered Ettore this way: "He was shy and timid; of sharp wit; with a lively sense of humor and an enormous human sensitivity... I was his youngest sister, and he loved me very much. He was so kind that he even did my math homework... I have many childhood memories of him. In the autumn we would go on holiday to Mount Etna. On moonless nights, Ettore would show me the sky, the stars, the planets: each occasion was a little lesson in astronomy. His words come to my mind still today, whenever I look into the star-filled sky... I like to remember him this way, as he invites me to look up at the sky and teaches me to call by name the stars."

Their well-off mother Dorina was equipped with a strong personality, which she availed herself of in acting resolutely "for the good" of her children. Dedication of this type invariably creates deep ties. As often happens, or used to happen, in Italy, particularly in the South, the ties between parent and children can become excessively strong or constraining. Of four children (apart from Ettore) only two got married: does this mean that only they succeeded in leaving their mother's protective wings? We cannot speculate over such things. We know that statistics based on such small numbers have no value; and unfathomable are the facts, contingent or otherwise, which influence a human being's vital decisions.

But we can say that Ettore, who was unquestionably very close to his mother, was too critical and sensitive not to feel the pressures of his family relationships. And, as someone who was already so shy and introverted, he likely also felt the weight and constriction of every other role that had been imposed on him or attributed to him. Not just that of obedient son, but also of a research group member (and of scholar devoted to theoretical physics, and later, of university professor).

This was accentuated after his father's death in 1934. Indeed, when Ettore made his "by now inevitable" decision, his brother Luciano murmured: "If father were still here, this would not have happened."

4

The Man: Recollections of Friends

Allow me to suggest that we set aside
the exceptional talents of Ettore as a physicist,
to emphasize what of him could evoke
his complex human spirituality,
so extensive and enlightened...
GILBERTO BERNARDINI

The Tomato Merchant

From the Scuola Normale of Pisa on 9th May 1984, Gilberto Bernardini wrote: "Dear Recami: ... Thanks to your copy of the letter[1] from Ettore to his mother, I am grateful to have exhumed a buried memory of my scattered meetings with him during the period I spent in Berlin with Lise Meitner... In Bologna[2] I spoke with

---

[1] Letter from Leipzig (MF/L2) of 22nd January 1933.

[2] In Bologna, on 26th April 1984 there was a historic meeting organized by the Italian Physical Society, as we have already mentioned, on "Fifty Years of Physics of Weak Interactions". Guests of honor were Edoardo Amaldi, Milla Baldo Ceolin, Gilberto Bernardini, Nicola Cabibbo, Piero Caldirola, Carlo Castagnoli, Marcello Conversi, Giuseppe Fidecaro, Ettore Fiorini, Raoul Gatto, Alberto Gigli Berzolari, Luciano Maiani, Lucio Mezzetti, Giuseppe Occhialini, Emilio Picasso, Oreste Piccioni, Bruno Pontecorvo, Giampietro Puppi, Franco Rasetti, Bruno Rossi, Antonio Rostagni,

Bruno Pontecorvo and I think that he, better and more thoroughly than I, could write you the "letter of testimony regarding Ettore Majorana" that you have amicably requested. In discussing it with Bruno, some memories were revived; and, among these, that with Ettore I avoided talking about physics because anything I could say would have been for him insignificant. As happened to me later with Pauli, I considered it easier for me and less trivial for him to talk about, for example, how great it was to be born after Michaelangelo and Beethoven. — I have not yet received what you published in 1975. I do not know if you are preparing a new and different publication with Ettore's sister, Maria. If so, then allow me to suggest that you set aside the exceptional talents of Ettore as a physicist, to emphasize what of him could evoke his complex human spirituality, much more extensive and enlightened than some writers could imagine. This spirituality, also emotionally, is evident when you read *The life and work of Ettore Majorana* written by Edoardo Amaldi and published by the Accademia dei Lincei in 1966… — Gilberto Bernardini." The writing of Amaldi cited here is essentially what we have included in the appendix.

Ettore was *extremely sensitive* and introverted, but profoundly kind. His reserve, shyness,[3] and difficulty

---

Carlo Rubbia, Giorgio Salvini, Claudio Villi, Gleb Wataghin, Giancarlo Wick, Emilio Zavattini, and Antonino Zichichi.

[3] Even Hans Bethe (Nobel Prize, 1967), in Erice, on 27th April 1981 confirmed to us that he had been able to speak to Ettore only two or three

with human contact, rendered even harder by his intelligence, did not prevent him from being genuinely affectionate. He would help for example his schoolmates, going so far as to sit for an exam in the place of a fearful fellow student! And his severe criticism softened when the judgment concerned his friends.

In his first letter to Carrelli, Ettore even feels the need to mention *Sciuti in particular.* Sciuti (later a professor at the faculty of Engineering in Rome) had been one of his few students in Naples. Ettore took an interest in these students and seemed quite happy with them. On 2nd March 1938, in his last letter to Gentile (MG/N1 in Part II of this book: "Letters, documents, testimonies"), he wrote: "I am pleased with these students, some of whom [*among them, certainly, were the three bright, intelligent female students of his, known as "the three white slaves"*] seem determined to take physics seriously." And when he took the chalk in his hand his shyness disappeared and he was transformed, as his hand gracefully filled entire blackboards with physical and mathematical symbols. This was described before the discrete video camera of Bruno Russo (1990), and then in Naples (1998) and to us personally, by his former student Gilda Senatore [*later professor Gilda Senatore, widow of Cennamo: then a young (and beautiful) female student of his*]: but it was easy to imagine it.

---

times during his stay in Rome, due to Majorana's extreme shyness. Bethe also recalls having heard Segré speak very well of Majorana.

To Sciuti — at least according to what he remembered —, one day Majorana asked: *Do my lectures really interest you?* and he brought Sciuti a typewritten copy of a few lectures. The entire series of his handwritten[4] lecture notes passed into the hands of Gilda Senatore, Francesco Cennamo, Anthonio Carrelli, Edoardo Amaldi, and Gilberto Bernardini, then to the archives of the "Domus Galilaeana" in Pisa, and finally the surviving portion, which corresponded to only 10 lectures, was published in anastatic form (Bibliopolis, Naples, 1987). We will return to this subject in Chapter 5. The notes for his Inaugural Lecture were recovered by the author, and appear here (Ms/2) in Part II.

Few people know that Ettore had a cheerful disposition, at least until 1933 (the year in which Ettore spent several months in Leipzig with Werner Heisenberg). His sister Maria, most of all, remembered his jokes, the laughter, and playing ball in the hallway of their home. As a boy, along with some friends he climbed behind the steering wheel of the family car — one of the first in those days — and of course not knowing how to drive he ended up against a wall. The memory of this was preserved by a long scar on his hand. Also Eng. Enrico Volterra (a former professor at the University of Texas at Austin — now deceased) spoke to us of the great times

---

[4] Ettore was likely thinking of writing a book for students, as well as he had probably thought of a book while writing his *Volumetti* of theoretical physics (see Chapter 2). But the existence of typewritten versions has *not* been confirmed at all.

he had spent with Ettore at the bar "Il Faraglino"[5] in Rome, and the discussions they had engaged in on cultural subjects at the "Casina delle Rose" of Rome's Villa Borghese.

Ettore had moreover a rich sense of *humor*, which is abundantly confirmed both through his letters and through numerous anecdotal episodes: such as the following (also from Volterra, narrated to us in Austin, and then repeated to us in Rome by Segré). During his university studies, Ettore and his classmates were following a calculus lecture taught by Professor Francesco Severi. They "followed" the lecture in a manner of speaking, as they were sitting in the back of the classroom engaged in gossip and idle chatter. But Ettore quickly realized that the teacher had taken a wrong turn (out of forgetfulness) in demonstrating a theorem, perhaps the Bolzano–Weierstrass theorem. And in fact, after short time, Severi gets flustered. Then, the companions of Ettore, who from the beginning of the lecture had not been able to disguise his amusement over the blunder by this respected and talented professor, decide to play a dirty trick on him, shouting: "Professor, please call Majorana to the board". Not knowing which way to turn, Severi conceded: "Well then, let this Mr. Majorana come!" Driven from his seat by his prodding companions, Ettore finds himself standing in front of the board,

[5] On Via Montecatini, at the corner with Corso Umberto ("Via del Corso"), in Rome.

and improvises (but perhaps he had already thought of it) the demonstration of that not simple theorem.

But we will depart from the perhaps too slight realm of anecdotes, and return for a moment to Ettore's letters. The previously mentioned Eng. Gastone Piqué, who was Ettore's closest friend from high school (Liceo "Tasso" in Rome) until his penultimate year of Engineering, kindly authorized us[6] to reproduce what remained of the abundant correspondence they exchanged in the summer periods between 1925 and 1928. "Correspondence" he told us "of a purely fraternal and youthful nature. It reveals the jovial and caustic temperament of Ettore, which many later denied that he had." He added: "Ettore was of very modest, naturally honest, caring and generous character." Some passages shed light on the wit and ironic (also self-ironic) spirit of Ettore Majorana.

From his family's vacation home on Mount Etna he wrote:

> "Passopisciaro, 17.X.1927 — Dear Gastone: ...I have not written to you up to now because I do not like to rush, especially in certain things. You should know that I have engaged in the most scientific of pastimes: I do nothing and time passes just the same. Actually I am working on an incredible number of things, but being vile matters of thought, their weight has to be shortened...

---

[6] See letter T/1 of the *Letters*, in this book.

If no stroke befalls me, I'll be there in a few days. Nor must you believe that it is impossible that a stroke befalls me in the prime of my life; on the contrary, regard it as quite likely. In fact, from birth I've been a stubbornly immature genius; time and straw did not help and never will,[7] and nature would not be so evil as to have me die prematurely of arteriosclerosis.

But though vast and unfathomable is the sea of my contempt for the entire sub-lunar world, it is not without jubilation that I prepare myself to cross the threshold of the renowned room[7] in Via Montecatini, nor is it without trepidation that I'll drink the bitter chalice, till the last drop — Affectionately, Ettore."

And from Hotel Tamerìci ("Tamarisks"), in the famous spa town of Montecatini Bagni, he was writing two months before, with a much lighter tone:

"August 2nd [*1927*] — Dear Gastone. I stoically drank three drops of bitter water, and then another ten drops, and then another six glasses. I am waiting; may God help me…

The languor which pervades me induces in my soul the most tender of feelings. What gentleness in the respite of one who has sipped a liter and a quarter of purgative water! … The delicate charm of your beautiful

---

[7] An allusion to the use of straw to speed the ripening of fruit. After a few lines, he refers to the famous watering place of Via Montecatini street.

land and the people who inhabit it, along with a subtle sense of nostalgia, complete the enchantment.

I hope to meet you in three days, perhaps tomorrow if I come to reserve the rooms, and I am looking forward to that moment; because there is no greater joy than to see you for your — Ettore.

P.S.: Duty calls. Farewell."

Very enjoyable is the postcard that soon followed. *To Mr. Gastone Piqué — Via Buonarroti, 42 — Viareggio:* "August 4th [1927] — Dear Gastone: I will come perhaps tomorrow, in the afternoon. I'll look for you actively and methodically: 1st) at the street number 42; 2nd) at the black cat; 3rd) at the Felice lido; 4th) beneath those windows. — Love — Ettore." With it, Ettore announces his arrival to his friend in the seaside resort Viareggio, specifying the places where he hoped to meet him. Piqué explained: "(i) at 42 Vespucci street [*or Buonarroti?*], where I lived; (ii) at the "Black Cat" café, located in the west pinewood, which I habitually frequented; (iii) at the bathing establishment "Felice", where I and my family had a rented cabin; (iv) "beneath those windows", by this alluding to the home of a young girl whom I was courting at that time." Eng. Piqué does not say it, but we shall whisperedly add here — for the sake of History — that those windows belonged to the daughter of the writer Giovacchino Forzano.

Humor is also an integral part of the letters written by Ettore to his parents. From Leipzig on 6th May

1933, for example, after having returned to Germany, he writes to his mother: "Dear mother — I arrived happily at the scheduled time… — I had an excellent trip. A bit crowded until Brenner. From Brenner to Munich I had only the company of a blind drunk Neapolitan gentleman. He was headed to Bremen to sell tomatoes, and honored me with his complete trust because he immediately assumed that I was going to Leipzig for the same purpose…"

Political Leanings

As to Ettore's political inclinations, it has been said[8] that he harbored sympathies for the nascent Nazism. From the letters we have (for example, the one to Gentile, MG/L1, of 6th July 1933) this is not as evident as some would have us believe. As we shall see later, these sympathies were once again mitigated by his critical, and ironic, nature. In this regard, let us read now a passage from the letter dated 22nd January 1933 from the Institut für Theoretische Physik, Leipzig:

> "Dear mother: … I like it very much in Leipzig … It snows gently here often … The Institute of Physics is, along with several other places, in a *delightful* position, a

---

[8] By E. Segré (and confirmed by Amaldi: see Appendix), based primarily on a letter from Ettore to Segré. A letter thought to be lost at first, then published by Emilio Segré in February 1988 in "Storia Contemporanea", vol. 19, p. 107 (and which we have included in Part II of this volume).

bit out of place between the cemetery and the lunatic asylum … The internal political situation appears permanently catastrophic, but to me it does not seem to be of much interest to the people. On a train I noticed the stiffness of a "Reichswehr" officer, alone in the compartment with me, who could not lay something on his luggage rack or make even the slightest movement without banging his heels together. This rigidity was evidently determined by my presence, and in reality it seems that this exquisite yet sustained courtesy to foreigners is part of the Prussian soldier's spirit, because, while he would feel disgraced if he had not rushed to light a cigarette for me, his demeanor prevented me from exchanging a single word with him apart from the compulsory greetings."

What we have found, on one hand, is that — upon beginning his employment at the University of Naples in early January 1938 — Majorana: (1) exibited a certificate attesting his registration in the National Fascist Party *starting from* 31st July 1933; (2) attached to his matriculation form, for the role of full professor, *the only* photograph of him in which he appears wearing the party's badge; 3) on 19th January 1938, took the oath of loyalty to the King and the regime, as did all university academics of that time (only about 10 had the courage to refuse to take the oath, among them mathematician and senator Vito Volterra: By doing this, renouncing his post as professor).

On the other hand, we have an interesting testimony from the great physicist Rudolf Peierls, who declared that toward the end of 1932 (that is, before leaving for Germany) Ettore was profoundly anti-fascist. Peierls, in fact, wrote the following[9] from Oxford on 2nd July 1984, to Donatello Dubini:

> "Dear Mr. Dubini, — Your letter of 4.VI, addressed to Cambridge, has only now reached me. I was with Ettore Majorana at Fermi's Institute in Rome during the winter of 1932–33. He appeared to be an extraordinarily gifted physicist, a bit shy, *and genuinely opposed to fascism.* This was before he wrote his famous works on nuclear forces and neutrinos. I have followed these works with great interest, though I never saw him again, as far as I can remember. I heard of his disappearance already in 1938, but then heard nothing more about it. Of course, this saddened all of us physicists, but we knew too few

[9] "Sehr geehrter Herr Dubini, - Ihr Brief vom 4.VI, der nach Cambridge addressiert war, hat mich jetzt erreicht. Ich war mit Ettore Majorana in Fermis Institut in Rom zusammen, im Winter 1932–1933. Er erschien mir als ein ausserordentlich begabter Physiker, etwas schüchtern, *und dem Faschismus sehr entgegengesetzt.* Das war bevor er seine berühmten Arbeiten über Kernkräfte und über Neutrinos schrieb. Ich habe diese Arbeiten natürlich mit grossem Interesse verfolgt, aber ich habe ihn nicht mehr gesehen, soweit ich mich erinnere. — Über sein Verschwinden habe ich wohl schon 1938 gehört, aber ich weiss nicht mehr wann oder von wem. Natürlich hat das allen Physikern sehr leidgetan, aber wir wussten zu wenige Einzelheiten, um über die Ursachen zu spekulieren [...] Mit besten Grüssen — Ihr — Rudolf Peierls."

of the details to argue about the reasons [...] With my best regards. Yours, Rudolf Peierls."

In addition to this testimony, the wise and "conclusive" observations by Leonardo Sciascia are also worth citing. But we'll do this at the appropriate time.

About Majorana "the man", there is still so much — almost everything — to say. But without denying ourselves a more articulated idea — first — of Majorana "the scientist".

5

The Scientist

En science, nous devons nous intéresser aux choses, non aux personnes.
MADAME CURIE

The Genius

In his letter of 27th July 1938, to Mussolini urging further investigation (in parenthesis, Mussolini wrote on Majorana's file the phrase *I want him found*, but nothing much came of it), *His Excellency*[1] Fermi — after having declared that Majorana was, of all the scientists in the world, the one that had most impressed him — added: "Able at the same time to carry out bold hypotheses and to criticize his own work and that of others; a highly expert calculator and profound mathematician who, beneath the veil of figures and algorithms, never loses sight of the real essence of the physical problem, Ettore Majorana has to the highest degree that rare combination of skills that form a first class theoretical physicist. And indeed, in the few years that he has been practicing his activities, he has been able to attract the attention

[1] The title of *His Excellency* (SE) originated from his membership in the "Accademia d'Italia".

of scholars all over the world, who have recognized in him one of the greatest minds of our time. And the successive news of his disappearance has dismayed all those who see in him someone who still has much to contribute to the prestige of Italian Science." At other times, as we know, Fermi compared Ettore to Galileo and Newton; considering him superior to everyone, including himself. Fermi's regard for Ettore was such that (as Piero Caldirola told us) Bruno Pontecorvo once reproached Fermi for being "too humble" in front of Ettore.

Often, when alone, Fermi and Ettore would stop to discuss physics in front of a blackboard, and sometimes loud cries would be heard coming from the room. When there was a series of difficult calculations to perform (such as evaluation of definite integrals, solving differential equations, etc.), Ettore and Fermi would compete with one another: Fermi, armed with his ever-present slide rule, which he was skilled in using; Ettore, doing everything in his head, facing a wall in order to concentrate (the equivalent of the table under which he'd hid as a child). And when Fermi declared that he was ready, Ettore would issue his verdict. Much later, Fermi and his slide rule would compete against von Neumann and his electronic calculator, one of the first, which von Neumann had helped build.

The details of the first meeting of Majorana with Fermi shed light on some aspects of Ettore, scientific or otherwise. Although they are known since the time Segré narrated them, it is worthwhile to reread them attentively.

> "The first major work written by Fermi in Rome [*on some statistical properties of the atom*] ... is known today as the Thomas–Fermi method ... When Fermi found that to proceed he needed the solution to a nonlinear differential equation characterized by unusual boundary conditions, with his usual energy in a week of hard work he evaluated the solution with a small hand calculator. Majorana, who had entered the Institute just a short time earlier and who was always very skeptical, decided that Fermi's numerical solution was probably wrong and that it would be better to test it. He went home, changed Fermi's original equation into a Riccati-type equation, and solved it without the help of a calculator, using his extraordinary aptitude for numerical computation (he could have easily become an act in a variety show, displaying the ease with which he performed the most complicated arithmetic calculations from memory). When he returned to the Institute, with a skeptic's eye Majorana compared the piece of paper on which he had written his result, with that of Fermi's notebook, and when he found that the results coincided exactly he could not conceal his astonishment."

The calculations Ettore performed that evening, or that night, have been found by us, and recently studied by alert colleagues. What Fermi had failed to do (to solve that differential equation in, as it is said, *analytic* form), Majorana achieved in a few hours, and following two independent paths: The first took him to an Abel's equation (rather than Riccati's, as affirmed above by Segré); while the second was completely unknown to Mathematics. Our friend Salvatore Esposito, a talented theoretical physicist, who interpreted the pages containing these calculations, had to devote two months of intense work to this task. It seems that Fermi did not exaggerate in comparing Majorana to the greatest geniuses of physics.

Ettore himself, however, invited to prepare a curriculum vitae,[2] would write about himself with his usual modesty (and this in May 1932, when he had already completed his most important works): "*Information on my didactic career* — I was born in Catania, August 5th, 1906. I earned a high school diploma in classical studies in 1923; I then regularly attended engineering studies in Rome university until the beginning of the final year. In 1928, desiring to study pure science, I requested and obtained admittance to the Faculty of Physics, and in 1929 I graduated with a degree in Theoretical Physics under the direction of S. E. Fermi, writing my thesis on *Quantum theory of radioactive nuclei* [3] and obtaining

---

[2] Written at the insistence of Fermi, who secured a grant from the National Research Council (CNR) so that Ettore could spend around six months (from January 1933) in Leipzig and Copenhagen. See correspondence relating to this scholarship, in Chapter 9.

[3] His oral "papers" were entitled: (i) *On a photoelectric effect observed in audions*, (ii) *On the equilibrium configurations of a rotating fluid*, (iii) *On statistical correlations.*

full marks and honors. In the subsequent years, I freely attended the Institute of Physics in Rome following[4] the scientific movement and carrying out theoretical research of various nature. Uninterruptedly, I took advantage of the wise and inspiring guidance of S.E. Professor Enrico Fermi." By "Freely", he means *voluntarily*; that is, without an official position, without receiving a penny in salary (and so it would remain until November 1937). With regard to the situation faced by recent graduates conducting scientific research, the Italian University can boast of not having changed.

The Scientific Work of Ettore Majorana

"*En science, nous devons nous intéresser aux choses, non aux personnes* (In science, we must interest ourselves in things, not in people)," the Polish Marya Sklodowska Curie,[5] universally known as Madame Curie, was compelled to proclaim. We must fully transgress this rule, as we are interested in the whole personality of Majorana. Madame Curie's quote, however, does remind us that we cannot understand a scientist if we neglect his scientific work.

Ettore published few scientific articles — nine, like Beethoven's symphonies — besides the semi-popular writing "The value of statistical laws in physics and social sciences", which appeared posthumously in

[4] He writes, one should note, *following the:* and not *following its.*

[5] Nobel Prize, 1903 (for physics) and 1911 (for chemistry).

Scientia in 1942, edited by Giovannino Gentile, since Ettore had thrown it out. Note that Majorana went from engineering to physics in 1928 (the year in which he published an article, his first, written together with his friend G. Gentile), and then devoted himself to theoretical physics for a few years only. But Ettore also left us very many unpublished scientific manuscripts, mostly stored at the "Domus Galilaeana" in Pisa, of which, in collaboration with M. Baldo and R. Mignani, we compiled a preliminary catalog in 1987. An analysis of these manuscripts enables us to confirm how: (1) as a physicist, Ettore was extremely diligent and accurate in his work. All of his discoveries were preceded by a long and tireless series of "calculations", done and redone (even for a genius, science cannot be a simple game of intuition, as his legend would seem to suggest); (2) within his unpublished material, many of the hints and works are still relevant and topical: We, together with the mentioned colleagues, went through this material and discovered that hundreds or thousands of pages[6] are still quite original and significant for contemporary physics, but not all of them at the least have thus far been "interpreted" and published by us[7]; (3) all the existing material seems

[6] Copy of a few hundreds of such pages having been transmitted by the present author, long ago, to the *Center for History of Physics* of the American Physical Society, for their "Niels Bohr Library".

[7] Not to mention old papers like "About a Dirac-like equation for the photon, according to Ettore Majorana" (*Lettere Nuovo Cimento*, vol. 11, 1974), let us recall two Books of ours reporting Ettore's lecture notes,

to have been written before 1933 (even the draft of his last scientific paper, "Symmetric theory of electrons and positrons", which Ettore would publish on the threshold of the competition for the professorship in 1937, appears to have been ready since 1933, when the discovery of the positron was indeed confirmed); (4) little is known of his activities in the next few years (1934–1938), apart from a series of 34 letters of response from Ettore (written specifically from 17th March 1931 to 11th November 1937) to his uncle Quirino, who urged him to provide a theoretical explanation for the results of his experiments. These letters, received initially by Franco Bassani and by us, courtesy of Silvia Majorana Toniolo, are mainly of a technical nature; so much so that we could publish only a small part of them in the Second Part of this book [all of them have been published by others at Bologna in 2006]. But they reveal that, even in those years, at least for the sake of his uncle, Ettore knew how to return to physics, demonstrating that he had not lost the skills of a master theoretician. And about this we shall meet stronger confirmation.

"E.Majorana — Lezioni all'Università di Napoli" (Bibliopolis, Naples, 1987, and 2006); and two Volumes, of about 500 pages each, presenting scientific manuscripts left unpublished by Majorana: namely, his *Volumetti* ("E.Majorana's Notes on Theoretical Physics", Kluwer, 2003, and Zanichelli, 2006), and a selection of his *Quaderni* ("E.Majorana's Unpublished Research Notes on Theoretical Physics", Springer, 2009). Another Volume, discussing several Majorana's (known or unknown) scientific achievements, has recently appeared c/o the Cambridge University Press.

Indeed, his sister Maria remembered that Ettore — who had begun to visit the Institute less and less frequently, starting from the end of 1933, that is, since his return from Leipzig — continued to study and work at home many hours a day, and at night, also at the time. Did Ettore study solely literature and philosophy (he especially loved Pirandello, Shakespeare, and Schopenhauer), or "game theory"[8] and naval strategy (his passion since childhood), as well as economics, politics, and medicine, or did he continue to devote himself to physics? From his letter to Quirino, *MQ*/R14 of 16th January 1936, we have a preliminary response, because we learn that Ettore *was engaged* "for some time on quantum electrodynamics". Considering Ettore's modesty when expressing himself, this means that during 1935 Majorana had profoundly dedicated himself to original research in the field of quantum electrodynamics, at the very least. And again in 1938, in Naples, Carrelli would have the impression that Ettore was working on something important, which Majorana did not wish to talk about.

What happened to the notes, writings, and articles associated with all this activity? Bruno Russo was the first to receive a testimony which makes "our veins and pulses quiver". But first we must examine what Majorana *published.*

---

[8] The study of strategic decision-making.

His First Publications

Let's return, then, to his published articles. The first ones, written between 1928 and 1931, relate to problems of atomic and molecular physics; mostly matters of atomic spectroscopy or chemical bonding (always within the realm, of course, of quantum mechanics). As E. Amaldi writes, a thorough examination of these works leaves one struck by their high quality: They reveal both a deep knowledge of the experimental data in even the most minute detail, and an uncommon ease, especially for that time, in exploiting the symmetry properties of the "quantum states" to simplify the problems qualitatively and to choose afterward the most appropriate way for a quantitative resolution. A recent paper, by Arimondo *et al.*, appeared in an important Journal of physics (see the Bibliography), stressed that in two 1931 Majorana's papers there were introduced the *auto-ionization*, long before the known U. Fano's work. But let us fix our attention now on the following article.

Atomi orientati in campo magnetico variable (Oriented atoms in a variable magnetic field), appeared in the scientific journal of physics *Nuovo Cimento*, vol. 9 (1932), pp. 43–50. This article, famous among atomic physicists, introduces the phenomenon now known as the Majorana–Brossel effect. In the article, Ettore predicts and calculates the change in the shape of the spectral lines due to an oscillating magnetic field, which was partially linked to an experiment attempted in Florence a few years earlier (albeit unsuccessfully) by G. Bernardini and E. Fermi. This work has remained a classic in the treatment of "non-adiabatic" spin reversal, or *spin-flip,* processes. His results, once extended, as suggested by Majorana himself, by Rabi[9] in 1937 and then, in 1945, by Bloch[10] and Rabi (who, both Nobel laureates, helped to spread what Ettore had discovered 13 years prior), formed the theoretical basis for the experimental method also used to reverse the *spin* of neutrons by a radiofrequency field: A method employed still today, for example, in all polarized neutron spectrometers. This article also introduces the so-called "Majorana sphere" (to represent spinors by using a set of points on a spherical surface), about which Roger Penrose — for instance — has spoken enthusiastically in his recent books (see Bibliography).

Ettore's last three articles are all so important that none of them can be ignored.

The Infinite Components Equation

The article "Teoria relativistica di particelle con momento intrinseco arbitrario" (*Nuovo Cimento*, vol. 9 (1932), pp. 335–344) is a typical example of work which was so far ahead of its time that it would be understood and fully assessed only many years later.

At that time it was common belief that quantum equations compatible with Relativity (i.e. "relativistically

---

[9] Nobel Prize, 1944.

[10] Nobel Prize, 1952.

invariant") could be formulated only for particles with zero or half *spin*. Convinced of the contrary, Ettore began constructing suitable quantum-relativistic equations for the successive possible spin values (one, three-halves, etc.), even inventing a general method (equivalent to the one later published by Dirac, and by Fierz and Pauli) to determine the quantum-relativistic equation corresponding to any spin value. He then discovered that a *single* equation could be written to represent an infinite series of cases. That is, an entire infinite family of particles with any spin value (remember that the known particles, now numbering in the hundreds, at that time could be counted on one hand!). Thus, he ignored all the individual cases studied and no longer published them, in order to dedicate himself solely to these equations "of infinite components", but without neglecting the observation that they could describe not only ordinary particles, but even tachyons.

To carry out this program, he invented a technique for the "representation of a group" several years before the presentation of these techniques by Eugene Wigner (Nobel Prize, 1963). Moreover, Majorana conceived, and used for the first time, the unitary *infinite-dimensional representations* of the Lorentz group, which would be rediscovered by Wigner in 1939 and 1948. To grasp the importance of this, let's return to what Ettore himself wrote to his father from Leipzig on 18th February 1933: "My last article published in the *Nuovo Cimento* contains an important mathematical discovery, as I was able to confirm through a consultation with Professor van der Warden, a Dutch who teaches here and is one of the leading authorities on group theory".

The theory was reinvented by Soviet mathematicians (in particular Gelfand and his associates) in a series of articles from 1948 to 1958, and finally applied by physicists years later. Ettore's initial article, on the other hand, remained in the shadows for a good 34 years. That is, until Amaldi translated it and alerted American physicist D. Fradkin of its existence. Fradkin, in turn, amazed the high-energy physics theoreticians by finally making available in the public domain (in 1966)[11] what Majorana had achieved so many years before. From 1966 onward, Ettore's fame grew steadily also among elementary particle physicists.

Exchange Forces

When news of the experiments of Joliot-Curie[12] reached Rome at the beginning of 1932, Ettore realized that they had unknowingly discovered the "neutral proton". So, even before the official announcement of the discovery of the *neutron*, made shortly afterward by Chadwick,[13] Majorana was able to explain the structure and stability of atomic nuclei using protons and neutrons (his manuscripts reveal that he had already experimented with

---

[11] D. Fradkin, *American Journal of Physics*, vol. 34 (1966), p. 314.

[12] Nobel Prize, 1935 (for Chemistry).

[13] Nobel Prize, 1935 (for Physics).

this problem, though unsuccessfully, using protons and electrons — the only particles previously known). Ettore's work was thus the precursor to the pioneering work of D. Ivanenko. But he would not publish anything, nor allow Fermi to mention it in Paris in early July, as related by Segré and Amaldi (see also the Appendix). His colleagues recall that even before Easter, Ettore had come to the most important conclusions of his theory: that protons and neutrons were bound together by quantum forces arising simply from their *indistinguishability*, that is by forces due to the *exchange* of their respective spatial positions (and not of their spin too, as Heisenberg will do), so as to obtain the alpha-particle (and not the deuteron) as the saturated system with respect to the binding energy.

Only after Heisenberg had published his article on the same topic would Fermi be able to induce Majorana to go to Leipzig to work alongside his talented colleague.[14] Finally, Heisenberg was able to convince Ettore, although much too late, to publish his results in an article entitled "Über die Kerntheorie", which appeared on 3rd March 1933 in the *Zeitschrift für Physik*, vol. 82 (1933), pp. 137–145.

These nuclear "exchange forces" were called Heisenberg–Majorana forces. With great modesty, Ettore wrote to his father in the same letter cited earlier (dated 18th February 1933): "I wrote an article on the structure of

---

[14] Nobel Prize, 1932.

nuclei that Heisenberg liked very much, even though it contained some corrections to his theory." And a few days later, on 22nd February, to his mother: "In the last "colloquium", the weekly meeting attended by hundreds of physicists, mathematicians, chemists, etc., Heisenberg spoke of the theory of nuclei, and he had much praise for the work I've done here. We've become good friends…"; and later, on 14th September, informing the CNR: "Prof. Heisenberg has accepted the findings in my work and has given it wide exposure… in a general *report* on nuclear physics intended for a future international conference (and currently being printed)."

The value of this publication regarding the stability of nuclei was immediately recognized by the scientific community, and especially by nuclear physicists. As we know, this was a rare occurrence for Ettore's writings — and was also thanks to the timely "propaganda" of the famous Heisenberg, who had a few months earlier received the Nobel Prize.

Ettore's aversion to publishing his findings when in his hypercritical judgment they were not general enough or not expressed in a compelling and elegant mathematical form, became also an affectation for him. In Amaldi's words:

> "At times, while talking to a colleague, he would casually state that the previous evening he had performed a calculation or developed a theory concerning some unclear phenomenon that had captured his attention or the attention of any of us at that time. In the discussion that

> followed, Ettore would at some point laconically pull out a packet of Macedonia cigarettes from his pocket (he was a heavy smoker) on which he had written in tiny but neat handwriting the main formulas of his theory or a table of numerical results. He would copy some of the results on the chalkboard, as much as was necessary to clarify the problem, then end the discussion, smoke the last cigarette, crumple the packet in his hand and throw it into the waste basket."

Also of particular interest are two passages from another letter Ettore wrote from Leipzig on 14th February 1933, to his mother:

> "... The environment at the institute of physics is very pleasant. I am on excellent terms with Heisenberg, Hund and with all others. *I am writing some articles in German. The first is ready*, and I hope to eliminate some linguistic confusion during the proofreading."

The article that was "ready" was, of course, the one regarding the nuclear forces mentioned earlier, which would however remain *the only article* written in German. Again, in his letter of 18th February, he told his father: "...I will also extend and publish, in German, my last article published in the *Nuovo Cimento*." We will return to this point, but for now it is very important to know that Ettore was writing other works, because he published nothing more while he was in Germany nor after his return to Italy, apart from the article of 1937. Let us discuss this article in what follows.

The Majorana Neutrino

As we have mentioned, from the remaining manuscripts it appears that in those same years (1932–1933) Majorana formulated the essentials of his theory of symmetric electrons and anti-electrons. He would have formulated them, that is, as soon as news spread of the discovery of the anti-electron or "positron". But Ettore would publish his theory, titled "Teoria simmetrica dell'elettrone e del positrone", *Nuovo Cimento*, vol. 14 (1937), pp. 171–184, only much later, in the process of his participation in the competition for the professorship. This publication was initially noted almost exclusively for having introduced the famous Majorana Representation of "Dirac matrices" in real form. A consequence of this theory was that a neutral "fermion" can coincide with its own anti-particle. Ettore suggested that neutrinos could be particles of this type.

This theoretical work was quite important to Ettore. This was observed by Carrelli, who discussed it with Ettore during his brief period of lectures in Naples.

As with other writings of Majorana, this article began to gain popularity only 20 years later, beginning in 1957. Following this, his work enjoyed progressively increasing notoriety among particle physicists and relativistic field theory physicists.[15] Today, it is fashionable to use expressions like "Majorana spinors", "Majorana mass", or

[15] In 1981, for example, a Japanese physics magazine released this article (in English, with translation by Luciano Maiani), about 40 years after its initial publication.

"Majorana neutrinos", and some particles are occasionally called "majorons". And more recent experiments seem to indicate something of immense scientific importance, if verified — that neutrinos be just as predicted by Majorana, and not of the type codified by the famous Dirac.

This last article published in 1937 by Majorana has been so much re-evaluated, seventy years after its appearance, that we owe the reader the following information. In this paper, besides the suggestion of the "Majorana neutrino", there were introduced the *Majorana fermions*, objects which coincide with their own anti-objects. Well, Majorana fermions have been actually discovered in a series of experimental works, published during the last ten years in the most prestigious Journals of physics (like Science, Nature, Physical Review Letters), as objects of condensed state physics. Moreover, theoretically and experimentally, they have been found to have essential importance even for the realization of quantum computers. At last, the very same article introduced also *Majorana algebras*, that have been already used by now — through their "Involutions" — for studying, e.g., the largest existing sporadic Group, called the Monster Group due to its mathematical complexity. In short, eighty years after its publication, all the most advised scholars in the world are recognizing that its author, Ettore Majorana, was really a genius "like Galileo ad Newton" as declared in due time by Enrico Fermi...

Despite this, Majorana's published work is still relatively unknown, in general and remains, for physics, a veritable gold mine. Recently, for example, Carlo Becchi noted that the first pages of the last Majorana's article contain also a clear formulation of the "quantum action principle", which in later years led to the most important developments in the so-called relativistic quantum field theory.

Testimonies of Colleagues

Many of Ettore's other ideas, those that did not remain solely in his mind, left traces only in his unpublished papers, or in the memories of his colleagues.

One of the most interesting accounts that we have collected is from Giancarlo Wick. From Pisa, on 16th October 1978, he wrote:

> "Dear Professor Recami: ... The scientific contact [*between myself and Ettore*] which Segré mentioned to you did not take place in Leipzig, but in Rome at the Volta Congress (thus well before Majorana's stay in Leipzig). The conversation took place at a restaurant, in the presence of Heitler, and therefore without chalkboard or written formulas. But, despite the absence of details, what Majorana described orally was a "theory of relativistic charged particles with zero spin based on the idea of quantization of fields" (second quantization). When much later I saw the work of Pauli[16] and Weisskopf, I was absolutely convinced that what Majorana had described was the same thing. Of course, Majorana published nothing and

[16] Nobel Prize, 1945.

probably did not speak to many people. I have absolutely no reason to think that Pauli and Weisskopf knew nothing about it ... — Sincerely yours — GC Wick."

And from M.I.T. (Cambridge, Mass.), on 16th May 1984, Victor Weisskopf wrote: "Dear Dr. Recami: ... I am very glad that you have found a letter in which Majorana says that he had good relations with me... I have only a vague recollection that I did have a discussion [*in Copenhagen, in 1933*] with Majorana about the newest developments in quantum electrodynamics."

The article by Pauli and Weisskopf that Giancarlo Wick refers to was released in 1934 (*Helvetica Physica Acta*, vol. 7 (1934), p. 709). Wick continued: "...I never had the opportunity later to speak to Heitler of this episode...one should not be surprised if he had forgotten it, because Majorana had spoken of the matter with that detached and ironic tone that he often used even regarding his own matters. In short, without giving them importance..."

Another testimony came to us, albeit indirectly, from the great but tragic figure of Bruno Touschek. On 29th October 1976, Eliano Pessa wrote from Rieti: "... We discussed with Touschek your work on Majorana in *Scientia* vol. 110 (1975) p. 577. He made some remarks about your list of Majorana's scientific works on page 585. In his view, you should add the theory of "The Majorana oscillator", which is implicitly contained in his theory of the neutrino. The Majorana oscillator is described by an equation similar to $q + \omega^2 q = \in \cdot \delta(\mathrm{t})$, where ∈ is a constant and δ is the Dirac delta function. According to Touschek, the properties of this oscillator are of considerable interest, especially as regards the energy spectrum. There is not, however, a bibliography for this…" The equation, then, is at *our* complete disposal. Are any readers interested?[17]

Nuclear Energy: Did Ettore Know?

We know that Ettore was a pure theorist, lacking in experimental inclinations. His lecture notes for the course held in Naples on 13th January 1938, for example, began this way:

> "In this first introductory lecture, I will briefly discuss the aims of modern physics and the meaning behind its methods, above all in as much as they contain more of the unexpected and original, compared to classical physics.
>
> Atomic physics, which will be our main topic, remains primarily a science of enormous *speculative* interest due to the complexity of its research, which reaches to the deepest roots of natural phenomena… This is in spite of its numerous and important practical applications, and the wider, perhaps revolutionary significance that its future may have for us."

---

[17] The problem, in truth, is not so much to solve the equation (which is well known), as much as to find out what Bruno Touschek had in mind (such as *boundary conditions*, for example): cf. Letter T/*11*, Chapter 11.

So while his interest in atomic physics is primarily speculative, Ettore also mentions the practical, *perhaps revolutionary*, applications that its future may bring. Had he already understood that the construction of nuclear weapons was close at hand?

Ettore returned from Leipzig on 5th July 1933. He resumed contact with the Institute of Physics, but it was increasingly less frequent.

But it was only in March 1934 (the year Ettore's father died, Pirandello won the Nobel Prize for Literature, and the Rome group obtained the "miraculous" experimental results in nuclear physics) that Fermi and his collaborators — Amaldi, D'Agostino, Rasetti, Segré and later, Pontecorvo — began to explore the idea of producing artificial radioactive elements by bombarding with neutrons.[18]

The neutron appeared to be the ideal nuclear "bullet" to produce the transmutation of elements because it does not have an electric charge and is thus not affected by the electrical repulsion of the nucleus. The obvious, and common, belief was that more energetic neutrons were more efficient at inducing reactions in the target nuclei. Fermi, on the contrary, soon became aware of the unexpected

---

[18] The forward-looking Professors G. Trabacchi and D. Marotta, the latter the Director of the Istituto Superiore di Sanità, gave them as a "source" *one gram* of radium: an amount with an enormous value. When, much more recently, attempts were made to prevent Italy from freeing itself from energy protection and from the "colonization" of technology, they would

fact that *slow* neutrons are hundreds of times more effective than fast ones. This led to the famed "fountain of red fishes" experiments behind the Physics Institute in Rome, where the water served to slow down the neutron bullets. Among the many targets utilized, even the last natural element, uranium, was used in the hope that the absorption of a neutron by the nucleus of uranium would lead (via "beta decay", that is, the emission of an electron) to the formation of artificial transuranic elements. And this is precisely how the results were interpreted: but erroneously so. That an entire generation of physicists was unprepared to grasp how neutrons with the lowest energy succeeded, where the most powerful bullets failed, to cause *nuclear fission* (splitting) of uranium. The few voices calling for caution (such as that from Freiburg of the Czechoslovak Ida Noddack) went unheeded.

So, in 1934 the Rome group had produced nuclear fission.

But recognition of this fact came only at the end of 1938, with the experiments of Hahn and Strassmann. More precisely, it was Hahn's collaborator, Lise Meitner and her nephew Otto Frisch, who would be eventually dazzled by the idea that these experiments (and the ones in Rome in 1934) had produced uranium fission: "The exact explanation was given," wrote Otto Hahn, Nobel Prize winner in 1944 for chemistry, "by Meitner and

---

strike at the most far-sighted people and organizations: D. Marotta (I.S.S.) and F. Ippolito (C.N.E.N.).

Frisch, to whom we had given our written results before publication. They then communicated their findings both to us and to Niels Bohr, prior to the publication of their article." Bohr was leaving for the United States (by sea, as this was before the advent of airliners) in the company of Rosenfeld, to discuss it with their American colleagues, who immediately confirmed the reality of fission. The news became public at the end of January 1939.

The target nucleus of uranium, in dividing itself into two nearly equal parts (that is, into nuclei of elements with an atomic number of about half that of uranium), release an enormous amount of energy. This energy is a hundred times greater than that which is released in common nuclear processes, and a million times greater than what is released in typical chemical processes, such as combustion. Only a correct interpretation of the 1934 experiments could have enabled Ettore to understand what an enormous source of energy atomic nuclei could be, but this had not yet been understood when Ettore disappeared. It was only in 1939 that research aimed at making practical use of nuclear energy began. This research consisted of attempts to produce the nearly simultaneous fission of large numbers of nuclei, which in 1942 would lead Enrico Fermi (the new Prometheus) to build the first "atomic pile" in Chicago, and soon thereafter, unfortunately, the A-bomb.

Had Ettore already foreseen this in 1934? Did this contribute to his estrangement (which had already begun by late 1933) from Fermi's group? This could certainly be the case, though we have no proof. On occasion, Ettore declared that "physics was (or physicists were) on the wrong path." His sister Maria also reminded us of this. But we believe he was referring to theoretical or "speculative" issues, rather than ethical ones.

Moreover, if indeed Majorana had feared the liberation of nuclear energy, he would have also understood that he could be more useful to its cause alive than dead.

Do Other Unpublished Manuscripts Exist?

Let us return to the letter of 18th February to his father, where we find the surprising statement: "I will also extend and publish, in German, my last article published in the *Nuovo Cimento*." This project, as we know, never came to fruition, but it is worth noting that Ettore intended to generalize the work (no. 7) where he had introduced his infinite components equation. Indeed, the issue becomes quite important when one reads the letters he wrote in that period to the CNR. In the first (21st January 1933), Ettore states: "I look forward now to the elaboration of a theory for the description of particles with arbitrary intrinsic momentum, which I started in Italy, and regarding which I have given a summary report in the *Nuovo Cimento* (currently being printed)..." In the second (3rd March 1933), referring to the same work he even states: "*I have sent a paper on the theory of nuclei to the*

Zeitschrift für Physik. *A manuscript of a new theory of elementary particles is ready and I will send it to the same magazine within a few days…*" If we remember that the article, considered "a summary report" of a new theory, was already at an extremely high level, we can begin to understand of what great interest it would be to discover a copy of the *full* theory, which in March 1933 had already taken the form of a completed manuscript, and was perhaps already typewritten in German.

But Ettore did nothing more, such that in his final report (14th September 1933) to the CNR he did not even mention it; it had become taboo. After mentioning the article on the "Theory of nuclei", in fact, Majorana immediately shifted to speaking of research initiated in his *second* period in Leipzig: "*In the last period of my residence in Leipzig I started other work which, for health reasons, I was neither able to complete nor to approach a conclusion. I believe it would be useless to discuss it.*" Why? Why did he do nothing? One might suppose that *at the last moment* he found a serious error which invalidated his new theory. But knowing Majorana, this is not likely. Instead, we tend toward another possible explanation: the "referee" of the German magazine may have rejected his manuscript. It was so groundbreaking that he may not have understood it (unfortunately, most of the archives of the *Zeitschrift für Physik* from those years appear to have been lost during World War II), and Ettore was not one to battle with fools. Then, final blow could have come from the bureaucrats at the CNR, who would demand that Majorana's articles be released only in the then still provincial Italian-language magazines,[19] rather than confer prestige to the best international journals of physics. Ettore responded in kind (on 5th September 1933). But then he may ultimately have succumbed to the frustration and discomfort that a true genius feels when confronted with human stupidity which, althought affectionate as he was with others, must have manifested even more strongly in him than in ordinary mortals.

Let's not forget, however, that the previously cited letter of 16th January 1936 to Quirino revealed that Ettore subsequently continued to work in theoretical physics, focusing deeply (but perhaps not exclusively) on quantum electrodynamics. But where are the notes, writings, and articles related to all this activity?

A Televised Testimony

Therefore, there are other scientific manuscripts from Majorana that are unknown to us. When Ettore, for his recondite reasons, reached the point of abandoning physics, his family, and perhaps his life, what did he do with the results of the work he had done during his last four years?

---

[19] Later on, the situation changed, and *Il Nuovo Cimento*, for instance, became an international Journal, entirely written in English. Even if bureaucrats (of the Government, this time) went on putting its existence at risk, periodically conceding or negating a support that — even if poor — was important for the Italian journals of physics.

The motivations behind his abandonment were not impulsive, but carefully considered and purified. Ettore could not have been disrespectful of the conquests of human intelligence and toil; nor could he have ignored the value of the discoveries that had been made. He would have thought about what the consequences would be if he left his most important unpublished notes and writings in his hotel room, or his office at the university, and he did not like it. Just as he had decided no longer to publish the fruits of his thought, he would neither have wanted them deposited in an Institute, University, or Academy — directly or indirectly.

Ettore chose to entrust his papers (some, at least) not to the bureaucracy or Institutions, but to someone who to him simply represented "life": that beautiful young student with a strong personality, for whom he had perhaps harbored feelings. The day before leaving, he appeared in the doorway of the lecture hall and called out "Miss Senatore!" Then, drawing back a step in the gloom of the corridor, he said "Take these papers: later we'll discuss them." He opened and closed, as usual, without giving her time to speak: "But professor…". "Take them", he said, "and then we'll talk." And he left.

But Miss Gilda Senatore (later Professor Senatore) could not avoid showing them to her neo-colleague Francesco Cennamo, assistant to director Carrelli, when he became her husband. Dr. Cennamo then independently elected to show them to Carrelli, having no idea that Carrelli would sequester them (and it should be stated that some of them served as a starting point and inspiration, and *only* as such, for later writings by Carrelli). Gilda Senatore discovered this at the end of the World War II, and the thought of losing that priceless legacy was like a dagger in her heart. The folder contained, first of all, the notes for the 16 lectures Majorana had prepared, carefully written for the students he adored, and perhaps intended for use in producing a textbook. But does giving them to Gilda Senatore not leave the impression that Ettore had prepared the notes so carefully specifically for her? Of this we cannot be certain. But we are convinced that Ettore loved teaching. During the academic years of 1933–1936, in fact, Majorana had requested of the Institute of Physics to hold free courses; "free" as allowed by his possession of the "Libera Docenza" degree (qualification for university teaching); on advanced topics: The role of group theory in theoretical physics (see the first line of the penultimate paragraph of his first letter to Giovannino Gentile, written in December 1929!); Mathematical methods for atomic and molecular physics; and so on. These courses were approved by Corbino, but they do not appear to have been conducted. Could it be that none of the students understood the value of those topics? We owe this information to Alberto De Gregorio, who in 2005 acquired it from documents that had since lain dormant in the archives of the University of Naples, and which had previously escaped attention.

They urge us toward the belief that Ettore's participation in the competition for the professorships in 1937 was, in large part, motivated by his purely pedagogical desire to have at least a few young students participate in his profound and original vision of physical science.

However, the folder, along with the lectures, contained also incomplete notes, completed papers, and articles. There is reason to believe that they represented the results of at least some of the work that Majorana did in isolation between 1933 and 1938. Events dictated that professor Ms. Senatore would be denied the path Ettore had chosen for her. If she had been able to carry out his decision, we would now have the complete folder of manuscripts — *now*, that we would be capable of fully appreciating them. Carrelli's confiscation of the documents, which the hierarchical system made possible, changed the path of destiny and perhaps resulted in some of the most important manuscripts being lost. An incalculable loss, one which perhaps even Ettore could not have calculated.

This information came from professor Senatore herself, not only via the said testimony collected by Bruno Russo, but also as a result of her intervention during celebrations organized by the University of Naples in 1998, to commemorate the 60th anniversary of Majorana's disappearance, as well as from her live (very live) voice. She passed away only recently.

Of the writings contained in the folder, only his lecture notes were initially found. In fact, as mentioned at the beginning of Chapter 4, only 10 lectures. Till

2005, when our Salvatore Esposito, in collaboration with colleague Antonio Drago, found the notes for the remaining six lectures, taken by a very careful student. So, the previously mentioned volume in the Bibliopolis of Naples (1987) would be reborn in its new form: as the complete set of Majorana's "lectures".

Erice

Although widely known, we cannot fail to mention that Ettore's name has been memorialized since 1963 in the title of the institute "Centro di Cultura Scientifica E. Majorana" in Erice, near Trapani. The center was founded and directed by Antonino Zichichi, with Miss Maria Zaini and Drs. Alberto Gabriele and Pinola Savalli initially lending their industrious and consummate experience to its creation. The center is host to five or six dozen international conferences a year. Erice is an isolated, unspoiled, charming town of Elymian-Punic origin, located not far from the Greek masterpieces of Segesta and Selinunte. The atmosphere of the Ettore Majorana Center has been likened by some to that of Copenhagen in the 1920s. With the reputation it has established, it has helped keep the name of Ettore alive among scholars of hundreds of countries. There is also a small but growing museum dedicated in part to E. Majorana, to which we donated (together with Maria Majorana, Ettore's sister) several pages of his scientific manuscripts.

6

The Letters of Ettore Majorana

To see people's eyes on me, I felt a terrible sense
of oppression thinking that all those
eyes attributed to me an image that
was certainly not the one I knew, but
another that I could neither know nor prevent.
LUIGI PIRANDELLO *(Uno, nessuno e centomila, 1926)*

Rediscovery of the Letters

In March 1972, in collaboration with Ettore's sister Maria, we rediscovered at her residence in Rome a series of letters written by Ettore. *Rediscovered*, in the literal sense: to "discover again". These letters, in fact, had been hidden away in a safe by Ettore's brothers, first because those represented painful reminders to Ettore's mother, and also to preserve the family's privacy. Even Ettore's final letters to Carrelli were immediately delivered to Salvatore, and Carrelli would later claim to have lost them. Even Amaldi could report what they contained based solely on "hearsay", and in greatly altered form.

But for us, and for Ettore's sister, the letters were new. Not rediscovered, but discovered, and the emotion they provoked is still alive.

Subsequently, we began studying the handwritten papers filed at the "Domus". And we received, to cite one example, copies of letters written to Gastone Piqué and Giovannino Gentile, and several documents filed at the Central State Archive of Rome and at the CNR. Almost all of the letters (and photos), like most of the evidence, were tracked down and collected by us personally[1] beginning in 1969–1970. But we cannot avoid some occasional overlap with the documents published by Leonardo Sciascia. In fact, with Maria's permission, we provided copies of much of Ettore's correspondence to Professor Sciascia around 1973, part of which he used for his sharp essay in October 1975.

To avoid violating a long-standing tradition of family discretion, and even more out of respect for this great mind and man, at the time we published quite little of the material retrieved.[2] Only a few months prior to the release of Sciascia's book, we sent to director Arrigo Levi a couple of articles (which appeared in Turin's *La Stampa* on 1st and 29th June 1975, and then, joined in a suitable form together, in *Scientia*).[3]

---

[1] Except for the material listed in points (i)–(iv) of the *Acknowledgements*. We take this opportunity to point out that the two letters to Gentile of 1930 are filed at the Gentile Foundation of Rome.

[2] Apart from a brief interview released to G.C. Graziosi in 1972 (*Domenica del Corriere*, 28.11.72) and some documents passed in 1973 to E. Macorini e T. Chersi for the biographical work *Scienziati e Tecnologi* of EST-Mondadori (item *E. Majorana*, signed by Amaldi).

[3] *Scientia*, vol. 110, Milan, 1975, pp. 577–598.

Untill, in more recent years, once clamor and controversy were dormant, his sister (Maria) and nephews (Fabio, Ettore Jr., and Pietro Majorana) decided to release *all* of the available documents, as we have done and are doing.

The Historical and Scientific Importance of the Letters

The collection of Ettore's letters carries significant historical and scientific value, but for us it is also of great biographical and human interest, as we have noted in the writings addressed to Piqué (Chapter 4), and elsewhere.

We will focus our attention on the letters of 1933 (we'll analyze those of 1938 later). As we know, Ettore arrived in Leipzig on 19th January 1933, and on the 22nd he wrote:

> "Dear mother,... Everyone at the Institute of Physics has welcomed me quite cordially. I had a lengthy conversation with Heisenberg, who is extraordinarily polite and friendly. I am on good terms with everyone, especially with Inglis, the American I met in Rome and who now frequently keeps me company and acts as my guide. My German is improving visibly.
>
> In a few days I will receive a visit from Bernardini,[4] who resides in Berlin-Dahlem and has returned

[4] It refers to these words the letter sent to us by Gilberto Bernardini (see Chapter 4) on 9th May 1984, with which he is pleased that they have

temporarily to Italy. The weather is pleasant — somewhat colder than Rome, but without the wind... If any reprints of *Nuovo Cimento*[5] arrive, please send me only a few of them, ten at most, in an open registered envelope (stamped registered)... If Turillo [*his brother Salvatore*] happens to go to the Ministry after January 27th, he can pick up my decree of "Libera Docenza" (from Borsi)..."

On 7th February, he adds:

"Feenberg, another American physicist with whom I have become friends, has arrived from Rome. We get along quite well in German. Tomorrow the so-called "magnetic week" begins in Leipzig, which draws nearly all the physicists of Germany. I will see many old acquaintances again. I intend to remain in Leipzig until the end of February; March and April are, in fact, the vacation periods here. I'll probably make use of the opportunity to go to Zurich to visit Pauli, one of the most famous scientists alive today. Inglis will be with me,[6] as well as Bloch, a Swiss man I met here and who also possesses the virtue of speaking perfect Italian."

---

unearthed "a buried memory of my occasional encounters with Ettore during the period spent in Berlin with Lise Meitner".

[5] In all probability it refers to the article "Theory of relativistic particles with arbitrary intrinsic angular momentum", *Nuovo Cimento*, vol. 9 (1932) p. 335.

[6] Nobel Prize, 1952. As we remember, Heisenberg received the Nobel Prize the year before (1932); Pauli would receive it in 1945.

And at the end of the "magnetic week", on 14th February:

"Dear mother,... A quite animated international congress of physics took place in Leipzig. I established personal relationships with several famous people, particularly with Ehrenfest,[7] who obliged me to explain some of my work in minute detail and invited me to go to Holland.

On March 1st, I will go to Copenhagen rather than Zurich, as in Switzerland, like in Germany, schools are closed in March and April, while Denmark follows the Italian usage. In Copenhagen I will see Bohr [*Nobel Prize 1922*] and others with whom I am already personally acquainted. It is, together with Leipzig, the most important center for theoretical physics in Europe..."

A few days later, on the 18th, he confirms his plans to go to Copenhagen to his father:

"Dear father ... On March 1st I will go to Copenhagen to meet Bohr, the grandest inspiration of modern physics, though now a bit old and senile..."

Ettore's sarcasm regarding Bohr should not be surprising. For an expert mountain climber, who loves to take the most direct route, those who conquer

---

[7] Paul Ehrenfest, one of the founders of quantum mechanics, who died tragically (suicide) that same year 1933.

peaks by following natural trails or arriving by parachute are not worthy of much consideration. So Ettore, the refined, elegant rigorous climber of the peaks of theoretical physics, did not look sympathetically at colleagues who preferred to survey the general landscape of physics without too much attention to rigor. Ettore, moreover, speaks of Bohr without having met him in person, and was likely influenced by what he had heard from a gossip colleague. In any case, he rightly asserts that Bohr was the greatest inspirer of modern physics: it was the school of Bohr (which since 1913 had built the model of the atom known by all) in which almost all scientists involved in quantum mechanics were trained.

On 28th February, however, he declared to his father:

> "I'll probably stop in Leipzig for another two or three days, as I need to chat with Heisenberg. His company is irreplaceable and I want to take advantage of it for as long as he remains here.... in Copenhagen I have an old friend, Placzek, who spent a year in Rome last year."

He had already expressed similar sentiments to his mother on the 22nd:

> "I greatly regret having to leave Leipzig, where I enjoyed such a warm reception, and will gladly come back in two months ... [*Heisenberg and I*] became good friends after numerous discussions and several games of chess.

The opportunities for them come during the office hours he holds every Tuesday evening for teachers and students at the institute of theoretical physics..."

In any event, Ettore arrived in *København* on 4th March. From the guestbook of the Institute of Physics in Copenhagen, it appears that Ettore attended what is now called the "Niels Bohr Institutet"[8] from 5th March to 12th April 1933. After having met Bohr, as early as 7th March he wrote: "Dear mother,... Bohr is good-natured and likes the fact that I speak German worse than he, and is very concerned about finding me a guesthouse near the institute. I am on good terms with Møller and Weisskopf. Placzek has been invisible, as he has been busy writing his '*Handbuch*' since time immemorial. He phoned me several times, still speaking good trasteverino.[9] This evening I will have dinner at his house,[10] along with Weisskopf." The letter from Weisskopf that we encountered in Chapter 5 refers to these words from Ettore. Of course, more than 50 years later, Victor Weisskopf did not remember much ("I must admit that my memory is not good enough to remember the details of the meeting I had with him...").

---

[8] Located, as today, in 15–17 Blegdamsvej (pronunciare *Plàidams-vai*).

[9] The dialect of Trastevere, a district in Rome.

[10] Of George Placzek, Segré said: "Placzek also had extensive interests outside of physics, knew many languages, was learned in literature and

On 18th March, Ettore added:

"Dear mother,... Bohr has left for about ten days. He is now in the mountains with Heisenberg to rest... In Copenhagen he is quite popular. The owner of a large brewery built, and offered him use of, a charming cottage that one accesses by passing through mountains of beer barrels. It is notoriously difficult to find for those who go there for the first time. I went there once for tea. Bohr himself guided my steps, as I was fortunate enough to meet him as he was taking a leisurely bicycle ride around the area..."

And on 29th March: "...Bohr has returned. A new acquisition: Rosenfeld. That is, the whole of theoretical physics in Belgium..." Once again, Ettore's tone is playful; we know that he was seduced by irony. And keep in mind that he was writing not to colleagues, but family members who existed outside the world of theoretical physics. His attitude should thus be considered more irreverent and mocking than sarcastic.

In 1964, Rosenfeld wrote for Amaldi a short "letter of memories" about his meetings with Majorana: a letter which we have included in Part II.

After 12th April, Ettore returned briefly to Rome. (He'd already announced on 22nd February 1933: "I expect to

---

music, and had deep knowledge of history and politics. He was a man of wisdom; witty and with an honesty and strength of character difficult to equal."

leave Copenhagen a few days before Easter and go directly to Rome," and on 29th March wrote: "I do not need the issues of the *Nuovo Cimento* which I will find in Rome in about 15 days...I'll leave around April 12th").

In early May, he resumed writing home from *Leipzig*. On the 23rd, for example, he wrote to his mother: "The German situation is very quiet; mine in particular even more so. I have a pleasant relationship with Heisenberg who enjoys my conversation, and patiently teaches me German. I must use this language exclusively, after the departure of Bloch who was well-versed in Tuscan. The forthcoming departure of the 24 airplanes of Balbo for America has been announced with the great fanfare."

Biographical and Human Interest

Moving on from the scientific to the more properly historical and biographical aspects, some other passages from the letters of 1933 are worth reading.

Without forgetting, here, a postcard written when he was not yet 10 years old, on 19th July 1916: "Dear mother, we are all well and enjoying ourselves very much. Please send me the book 'La guerra sul mare' [The War on the Sea]. Greetings and kisses from — Your loving son — Ettore." A revealing postcard — his request for the book displays the interest Ettore had always shown for the naval fleets of various countries and their relative strength (in 1916, World War

I was in progress). The postcard was sent from Strada (Arezzo), probably during a trip organized for him, his brothers, friends, and classmates by the "Convitto Massimo" boarding school in Rome.

Immediately following his arrival in Leipzig, Ettore recounted to his mother a misfortune he endured on the border with Austria as a result of the type of currency he was carrying, a misfortune that caused Majorana to be momentarily detained by the Austrian police: "I arrived late last night after an excellent trip. I had to stop for a few hours in Kufstein due to the chaotic manner in which Austrian currency control operates. At Brenner they did not register the checks made out in my name, and I was reproached in Kufstein for not having reported them. In the end, the episode was clarified before the police chief of Kufstein and they did not give me any more difficulties. I learned this is a rather frequent event. I quite like it here. The city is beautiful and comfortable, its inhabitants very polite."

Two days later, he added: "Life here is not expensive and the many cafes and night clubs are also cheap, with great music and carnival-like entertainment, and are very crowded on Saturday night. — I often go to the cinema in order to accustom myself to German conversation... Many abandon themselves without restraint to roaring laughter; a sign, among many, of the meager social subjugation characteristic of the Nordic people..."

And on 7th March, now in Denmark, he switches to politics: "Dear mother, I have not spoken of the political situation in Germany, as it had not seemed very serious. There has been some shooting at night even in Leipzig resulting from electoral agitation, but no movement of a general nature. The elections are also forthcoming in Denmark. Huge processions of communists march through the center of the city chanting and displaying placards, mainly against Mussolini and Hitler. They cause more laughter than consternation." Rudolf Peierls testified that at the end of 1932 Ettore was "very opposed to fascism". From this letter, it does not appear so. But this contradiction may be explained, in part, by the fact that Ettore: (1) did not want to alarm his family, and (2) was perhaps simply using with its family the style of language that was common in Italy at that time. Certainly Ettore, as with the great majority of Germans, did not realize what was happening in Germany. This was exactly in 1933, when Hitler assumed full power.

This seems to be confirmed by the now famous letter Majorana wrote to Emilio Segré on 22nd May, 1933, also from Leipzig, and which was finally published by Segré in 1988, 50 years after Ettore's disappearance. We provide this letter in Part II. But we do not totally agree with a *naive* interpretation of its contents. Speaking of the warning signs of German antipathy toward the Jewish population, Ettore embarked on a cold and rational account of the events, without expressing clear sentiments of approval. It appears in it, rather, a personal grudge against Segré. In particular, this letter does not concede anything to the nascent Nazi racism. Clear proof of his condemnation of racism, on the other hand, comes from his own words in those days to his friend Gentile. On 7th June 1933, Majorana wrote:

> "... The internal situation in Germany is apparently as stable as in Italy, but you cannot make comparisons between the political maturities of the two populations. Germany, which cannot find within its culture and history elements sufficient to nourish a common sentiment among German-speaking peoples, is forced to resort to the foolish ideology of race which, apparently, has not adequately echoed in Austria. The struggle against the Jews, though partly justified by instinct, is not so for the reasons cited to support it; among these, sadly, the eternal issue of race dominates..."

There is a pronounced difference in the tone of the letter to Gentile compared to that of Segré, though written only two weeks apart. Why? In our humble opinion, the letter to Segré constitutes a bit of a reprisal by the 26 year old Majorana against his *Basilisk* colleague who, although his friend, likely did not forgo the opportunity to tease Ettore at times. It otherwise does not make sense for Ettore to send this letter to Segré, who was Jewish. Also revealing is a reference to "keeping one's mouths shut", which certainly makes sense if referring to a specific person, but not at all in the context of an entire community of individual people. In short, it was partially a boyish prank: but undoubtedly strong.

In any case, the final word on this point was had by Sciascia, whose irreproachable voice is certainly not suspicious:

> "It is 1933 and the anti-fascists in Italy can be found only in prison. Four years prior, there had been a "reconciliation" between the State and Church:... the bishops blessed the banners and proclaimed Mussolini a "man of Providence." The year before, even Pirandello had mounted the guard at the exhibition of the tenth anniversary of the "fascist revolution". Marconi presided over the Royal Academy of Italy, as Mussolini desired... The whole world admired Italian aviation enterprises. Academic and militant critics praised Mussolini's prose. At every speech by Mussolini, Piazza Venezia rumbled to a consensus that was echoed from the palaces to the slums. Soviet Russia participated in the Venice film festival... And should we ask of the twenty-six year old Ettore Majorana, who was disengaged from politics to the extent that you *could* be then disengaged, so distant, so closed within his thoughts, a net rejection of fascism, a harsh judgment on nascent Nazism? We must also bear in mind that letters from other countries were frequently opened and read..."

From Copenhagen, Denmark (29th March), he wrote with his usual tendency for caricature: "My health is still good. In Copenhagen one can find all the amenities desirable for material life. Its inhabitants have almost ceased to interest me. My stay of almost a month confirms that there is nothing left to discover of the soul of Denmark. The people are extraordinarily peaceful, almost passionless…" Previously (5th March 1933), he had observed: "Copenhagen is an enormous city with good architecture. The population, equally intelligent and civil from the highest to the lowest strata, is cut from the same template. This is undoubtedly the secret of the much-prized Nordic civility… In the long run it must be boring."

For our purposes, however, we are more interested in the final letters from Leipzig. At the end of June, the Majorana family (excluding Ettore's father) took a trip by car to Paris, and in July went to visit Ettore in Leipzig. Toward the end of July, Ettore resumed addressing his letters to Rome. Of particular significance are the last two letters to his mother. On 25th July, he wrote:

> "I am still under the doctor's care, with slow, but in his opinion, sure results.
>
> I have no intention of coming to Abbazia[11] because I would not be able to swim and the heat at the beach would be unbearable. I would like instead to go directly to Rome in a few weeks. If there are difficulties with

[11] A renowned seaside resort, now in Slovenian territory; which was the ritual summer destination of the Majorana family, headed by the mother.

> regard to service, laundry, delivery of keys, etc.., please list them and I will consider how to resolve them..."

And two days later, gradually abandoning his usual affectionate and controlled style, he devotes an entire letter to a rebuttal of her maternal insistence:

> "Dear mother, I received your letter, and later the one from Turillo. Your worries concerning my intention to go directly to Rome seem to me exaggerated. My health is already showing signs of marked improvement and I no longer have serious problems unless I cease to follow the doctor's prescription ... I think that I will feel better in Rome than in Abbazia. Also for meals... the ordinary choices in Rome are beyond all comparison... I do not understand why you say you want to come to Rome for a few days... You would bring me useless sorrow if you embark on so long and tiring a journey with no purpose or justification. But I do not intend to change my plans for the fear that you will carry out so unreasonable a threat."

So Ettore sees the "threat" from his mother (that she feels forced to return to Rome, at least for a few days) for what it is: a bit of blackmail.

The family also remembers another display of intolerance by Ettore. Preoccupied by thoughts and doubts that were far more serious, deep and important, following repeated discussions with his mother over her expressed desires for leaves, Ettore shouted: "And enough of this Paris!" A shout that made quite an impression, coming from him.

Digression

Leonardo Sciascia was one of few writers to devote an essay to a scientist, acknowledging in him a true human richness (although in doing so, he seems to have felt the need to put Ettore in opposition to other scientists). For over a century, most "cultured" people were not interested in science, nor did they realize that scientific knowledge was part of the ideal substrate of human life. This is surprising, when one considers that physicists and mathematicians, since the beginning of the twentieth century, have frequently taken on problems far more extensive than what was strictly within the confines of their science. One need only think of Einstein, Poincaré, Weyl, Heisenberg, Ehrenfest, de Broglie, Schrödinger, Bridgman; or Peano, Enriques, Cantor, Russell, Gödel, etc.

30 years ago, as Toraldo di Francia recalled, it was each time irrefutably necessary to endure those little wars ("that a physicist is not a biologist is certain, but that he is not a humanist is yet to be seen!") to get recognized to science the value of culture. This was, and is, particularly true in Italy. Why? As the famous epistemologist Mario Bunge writes in describing his relationship with Italian philosophy: "I had an elementary knowledge of it at a

young age when, admiring more than understanding it, I read Giordano Bruno. Shortly thereafter, I read Galileo with enthusiasm and benefit; and I read Croce with pleasure, perhaps even drawing value from it. I also read Campanella and Valla with great admiration. Telesio and Vico rather disappointed me (I was particularly struck by Vico's affirmation that the study of mathematics dries up the imagination: the idea that Vico had about mathematics evidently lacked *imagination*). Much later I learned to respect Aquinas (St.Thomas), but not to love him. The "contemporary" Italian philosophers from whom I most benefited, he writes in 1970, were Antonio Labriola, Federigo Enriques and Rodolfo Mondolfo. I am told that young people hardly read them anymore. How sad! Occasionally I re-read certain Italian philosophers, particularly St. Thomas, Bruno and Galileo. I was not, however, able to re-read Croce. After one has studied a bit of mathematics and a little science, the philosopher (not the historian) Croce becomes slightly more tolerable than Hegel, which is not much. His problems seem dead to me, and his method is to my mind more historical than philosophical... It is not easy to forget the pernicious influence exerted on contemporary Italian culture by his illogical and unscientific attitude..."

Until the nineteenth century, however, philosophy and science went hand-in-hand. In fact, they were often the product of the same minds (from Aristotle to Descartes, Pascal, Leibniz and Laplace). So what happened?

On the one hand, many of the same scientists isolated themselves, descending into their specializations. On the other hand, the magnitude of technological achievements led superficial observers to forget the true value of science as *knowledge*, leading them to confuse knowledge with its practical applications. On these two points, we will yield the floor to a man of letters and to a physicist. The Spanish philosopher Jose Ortega y Gasset described the specialist as a type of scientist "without historical parallel", who he considered a representative of the typical "uomo massa", part of the brute and ignorant mob that endangers true civilization: "This is a person who, among all the things that a cultured person has the duty to know, is familiar only with a particular science, nay, even of this science he knows only that small part in which he is engaged in research. He goes so far as to declare it a virtue not to deal with any of what remains outside the narrow domain that he himself cultivates, and denounces as *amateurish* the curiosity that aspires to the synthesis of all knowledge. It sometimes happens that, though segregated in the narrowness of his views, he actually manages to make new discoveries and to advance his science (which he barely knows), and with this, help to progress the whole of human thought, which he ignores with full determination…" Meditating on this, the physicist Erwin Schrödinger would recommend to his colleagues: "Never lose sight of the function of your particular field during the tragicomedy of human life. Stay in contact with life; not so much with practical life, as with the fundamental ideals of life, which are always much more important; and life will remain in contact with you ..."

On the second point, the same Schrödinger[12] wrote: "Admittedly, there was a period in recent times... which prompted one to undervalue the ideal mission of science. I place this period in the second half of the nineteenth century. That was an era of sudden and enormous scientific development, and together with it, an incredible and rapid development of industry and engineering. The latter have had such great influence on the material side of life that a great many forgot every other aspect..." (with this, he certainly did not mean to imply, as is clear, that material progress is not important: in Brazil, *when* part of the population was starving, there were perhaps 10 million abandoned children). To conclude — if some patience is left —, let us read the words of Sebastiano Timpanaro: "Our beautiful science, it is useless to disguise, has not yet managed to blend intimately with our culture and to become an essential element of it. Science is more or less widely studied in all schools, but our culture remains stubbornly philosophical and literary. This fact is due, among other things, to the same scientific progress that makes science inaccessible, or nearly so, to the uninitiated....; and

[12] Nobel Prize, 1933.

to the absence, in the domain of scientific history, of a brilliant critic comparable to a De Sanctis, and above all to the low regard for science of our principle philosophers, who are the real directors of our culture. All the theories about science supported by them, from those who claim that science is everything, to those who admit that it is only something, or less than that, are built on very few scientific notions about which the philosopher has just the vaguest information... Our philosophers are proudly sure of the theory of science, but this does not mean at all that they know all of science: they do not even know the elements, and they boast about it. Fortunately, the ever increasing importance that history is assuming, thanks also to these same philosophers; the discomfort that many now feel for the pure philosopher "who knows nothing about anything"; the ever stronger need that one feels for direct knowledge of the Classics of science, even within the scientific world, makes us feel that in the end a new humanism will triumph that is both classic and modern.... It leads us to hope that the time is near when we will finally begin to give our scientists the recognition they deserve, and therefore the chasm that has been artificially dug between science and classical culture will be filled." We hope so. Unfortunately, Timpanaro's words were written before most of us were born: in 1926.

7

Anatomy of a Disappearance

Who knows how many are like me,
in these same conditions , my brothers. One leaves
his hat and jacket, with a letter in his pocket, on the
parapet of a bridge over a river. And then, instead
of jumping, he quietly leaves, to
America or elsewhere.
LUIGI PIRANDELLO
(*Il fu Mattia Pascal*, 1904)

Before

An essential part of Part II is, of course, the letters written from Naples. An examination shows a sudden change in the tone of Majorana's final letters compared to the earlier ones. In the letters to his family, for example, Ettore's writing was generally well balanced (perhaps even "controlled", as stated previously), explanatory, witty, loving, and long. On 23rd February, a month before he disappeared, he said to his mother in a letter written from the Hotel "Bologna" in Naples: "Today they will give me a room in via Depretis, from which I will be able to watch the passage of Hitler *in three months*! Have you recovered from your two minor colds? I intend to come perhaps after Carnival. Loving regards — Ettore." On 12th January, in a letter thanking the Minister for the "highest distinction" conferred upon him by his appointment to full professor "out of competition", he wrote: "I wish to affirm that I will give all my energy to the school and to Italian science." Therefore, when on 22nd January he asked that his brother Luciano send him his share of their bank balance, there is reason to believe that at that moment he was thinking only of his own long-term accommodation in Naples. His desire to have his own residence is also supported by what he wrote to his brother Salvatore a week before his disappearance: "Naples, March 19th, 1938-XVI. Dear Turillo, …I will check whether it is possible to get a "'Health-book' for mother, but I do not see how I can confirm cohabitation with her, as I have an obligation to reside in Naples. In fact, I've already taken up temporary residence in the hotel in via Depretis, 72." And to us this does not appear to derive solely from a sense of duty to satisfy an obligation to live in the same city where he was teaching.

But that Saturday, 19th of March, having completed his internal struggle, Ettore had probably already made his "by now inevitable" decision. There was not "one grain of selfishness" in it, as if to say that for a considerable time, perhaps years, he had wondered to himself if he could morally come to this decision, or whether it was forbidden because it was based, at least in part, on selfish desires. Perhaps these thoughts had churned in his mind recurrently since 1934, until he convinced himself of the necessity to make such a decision. And as such (a necessity), he was cleansed of every "grain of selfishness".

And Ettore starts carrying out the project he had contemplated and agonized over for so long: to begin its "construction" (the words normally used in such cases, like "staging", do not apply in Ettore's case), most likely that same Saturday. Indeed, he sends a telegram to Rome to cancel his plans to spend Sunday at home, as he usually did. He then writes to his older brother, Turillo: "This time I am not coming because on Monday I have some things to take care of… I'm sending you a telegram so that you will not expect me tonight, but I will certainly come next Saturday."

Then a week of silence. The "next Saturday" (26th March) would be the day he wrote his last letter, from Palermo, to Carrelli.

On Friday the 25th, he again picks up his pen. In the letter (the first) to Carrelli, he realizes, he writes, "*the bother that my sudden disappearance will bring to you and the students.*" In the words of Sciascia, "Even the word *disappearance*, rather than 'death' or 'demise', we believe was used so that it would be understood as a euphemism; but it was not".

But in the same letter, in closing, he adds: "…*for whom I will hold dear memories, at least until eleven o'clock this evening, and possibly thereafter.*" This would have us believe that 11 o'clock is the hour of his intended suicide. It was, rather mundanely, the time of his departure — the ship was to leave at 10:30 that night. Taking into account a reasonable delay, 10:30 becomes 11 o'clock. At last, one could interpret his closing words ("and possibly thereafter") as a tribute to the hopes of religion. We believe the meaning could, again, be literal. But this is the same game "played at the limits of ambiguity" noticed by Sciascia's sensitivity.

During

Why, wondered Pirandello, when one thinks of killing oneself, he imagines himself dead; not dead to himself, but dead to others? Continuing with this line of thought, Vitangelo Moscarda[1] continues to dwell on his anguish with this question: "perhaps that was the moment to end it all, not so much to free myself from this torment, but rather to surprise those who envied me. …"

Ettore did not like to imagine himself dead, not even to others. Before leaving the hotel, prior to "setting sail", he left the following letter: "*To my family — I have just one wish: that you do not wear black. If you wish to mourn me then do so, but not for more than three days*". This is a concession to a social reality: the three days of full mourning in the Sicilian tradition, as we know. "*Afterward,*" he asks simply: "*keep my memory in your hearts.*"

Citing Pirandello on the subject of Majorana's disappearance is certainly nothing new. But it would be

---

[1] The protagonist of Luigi Pirandello's 1926 novel *One, No one and One Hundred Thousand.* ["Uno, nessuno e centomila"].

quite reasonable to imagine that on that same bedside table where he left the letter to his family, Ettore also kept Schopenhauer, Shakespeare ... and Pirandello. He, too, was a Sicilian: "glory", together with Empedecles, of Agrigento; so as, Bellini, Verga and now Majorana are the "glories" of Catania.

We still do not know why Majorana arrived at this decision, but the papers suggest how. How many times would Ettore have read and, more suggestively, seen at the theater *Il fu Mattia Pascal:* "...Who knows how many are like me, in these same conditions, my brothers. One leaves his hat and jacket, with a letter in his pocket, on the parapet of a bridge over a river. And then, instead of jumping, he quietly leaves, to America or elsewhere." And Ettore does exactly this. He knew that the most trivial things are the least believed. Except that Ettore puts his passport and money (which we calculated to be equivalent to at least $20,000 today) in his pocket, and he leaves the letter on the bedside table.

So Ettore leaves: by a ship; or rather by the *first* ship in his "project". Was his rekindled love for naval vessels during his final years nothing more than a mathematical interest in naval strategy and an engineer's attention to ship construction? Or was it an outward manifestation of an unspoken but already existing desire to escape to a destination far across the sea? At that time in Italy, those who emigrated dreamed of Argentina. The same

Mattia Pascal, after stating *America, or elsewhere*, specifies his thoughts by name: Buenos Aires.

Ettore vanishes, but he does not go *quietly*. Neither does Mattia. And Ettore is not Mattia Pascal.

During the night in the steamboat between Naples and Palermo, Ettore's mind and heart were not at rest, even if he was able to sleep. The police, his colleagues, and friends would believe him dead and would not look for him, just as he desired; this was the objective of the final letters he'd written in such poised and meditated handwriting. "Foreordained" says Sciascia. But once again he thinks about his family, his mother: Will they understand that he has left them some hope? His decision is a response to objective needs, thus to the domain of that which is right, ethical, and necessary. But won't the family's pain be too bitter? Once again, his doubts gain the upper hand; even in him who, when dealing not with human sentiments but with the calm and lofty things of nature, could construct breathtaking but stable architectures of thought; was able to "calculate" harmonious relationships with unsurpassed mastery.

Having just disembarked (apparently) in Palermo, he sends the urgent telegram we mentioned previously, and it was in Carrelli's hands that same morning at 11. Ettore knows that Carrelli, like most, had thought about suicide, and it is with a hint of bitter irony that he writes in the letter that followed: *the sea has refused me*. Perhaps for a moment he thinks of abandoning his plan, at great sacrifice, and returning[2]; not to the Institute, but home. That is, to the *Hotel Bologna*, which at that time was his home. Perhaps he indeed considers returning when he says: *I do, however, intend to renounce my teaching post.* This new, *additional* task of having to behave as others require would have weighed on him too heavily: to live like many others live, and expect *him* to live. The "others" are many, and include almost no one who meet him on *his* level. "Too much he was already concentrated in the horror," wrote Pirandello, "of being confined in a prison of any shape." And he defends himself: "*do not take me for an 'Ibsenian girl', because the case is different*".

In making his resolution the previous week, he had already taken the most difficult step. And he was able to look at his own situation with sufficient detachment to say: "the case is different." And he does not even say *my* case. It is just *a* case, that is different.

At this point, he realizes that he has left his family with tangible hope, and he can then proceed with his plan to escape from the old world, to renounce his ancient name. "It is nothing more than a headstone inscription, a name. It is necessary for the dead. For those whose lives have finished," Pirandello wrote. "I am alive and not finished. Life does not finish. And life does not care

[2] And he sent this telegram, probably, so that they will keep his room and will not go through his papers; so that they will neither find nor deliver the letter addressed *To my family*.

about names... I am this tree... tomorrow a book or the wind. The book I read, the wind that I drink. My all *out*, wanderer... Only like this can I live now, reborn from moment to moment, to prevent thoughts from working in me again, from building within me the emptiness of vane constructions." Of course, Vitangelo Moscarda does not exist on Ettore's level, and perhaps could not even imagine it. Nor should we forget the admonition of Enrico Fermi: "With his intelligence, once Majorana had decided to disappear (or to make his corpse disappear), he would have succeeded." But it is worth reading what one of Pirandello's critics, G. Croci, had to say:

> "Like Mattia Pascal, the living-dead of the eponymous novel, the protagonist of *One, No one, and One-hundred Thousand*, Vitangelo Moscarda, finds himself engaged in a desperate experiment: to rebuild a life that is free from the constraints imposed by nature and convention, and to assert his genuine character through an act of free choice."

As we were saying, Ettore indeed left his family with tangible hope. Yes, his mother would remain convinced that her son had *not* committed suicide ("Ah! When Ettore returns he will hear it from me, with all the troubles he's getting me!", she would occasionally exclaim), and this belief persevered throughout her life, such that she left him his share ("for when he returns") in her will.

After

"Your Excellency, I turn to you, the supreme inspirer and moderator of Justice, so that they may be intensified to the extent possible the measures most suitable for the recovery of my son, Ettore Majorana. He was a professor of theoretical physics at the University of Naples, a post to which he had been appointed for exceptional merit in November.

His painful and sudden disappearance has by now lasted for four months and there has been only one sure trace of him. In the last days of March or early April, Ettore Majorana appeared in a very agitated state at the Superior of the Church of Gesù' Nuovo, in Naples, and asked to be hosted in a retreat to experiment with religious life.

Having not been immediately accepted, for obvious reasons, he disappeared and there has been no news of him since. All of the investigations carried out by the Ecclesiastical Authority have been fruitless.

He was always wise and sensible, and the drama in his soul or nerves seems to be, in any case, a mystery. But one thing is certain, as all his friends, family, and I myself as his mother will attest: one never saw in him any moral or medical evidence that would suggest suicide. On the contrary, the serenity and seriousness of his life and his studies allow, indeed require, that he be considered solely a victim of science.

And of this there is no better witness than Royal Academy of Italy member S.E. Fermi, who was his mentor and friend and who wanted to refer Your

Excellency to the enclosed letter, as an expression of the esteem he has for my son.

It is my understanding that the police have diligently endeavored in the search, unfortunately, thus far without success.

If I am permitted an opinion, the search for my son should start in the countryside, in some farmer's house where it is far easier to escape the vigilance and careful searching of the police, and where he can conserve for quite some time the few thousand lire that he brought with him.

But so far there have been no reports, even though the search bulletin has been displayed three times. In the event that my son is abroad, I make known to Your Excellency that his passport (No. 194925) expires in August and must be renewed at a consulate.

Your Excellency, it is an illness caused by his worthy studies, perhaps completely curable but possibly destined to worsen beyond remedy if neglected. Your powerful intervention may determine the fate of this search and the life of a man.

Your Excellency, to whom we attribute the most ingenious and generous initiatives inspired by enlightened understanding and crowned by triumphant success, before you kneels a distraught mother, but one full of faithful hope."

This is what Ettore's mother wrote in a letter to Benito Mussolini from Rome on 27th July 1938. Fermi's letter, written the same day and for the same purpose, is the one we have already seen.

It is clear that his mother was expressing the thoughts of the entire family, and among the possible hypotheses, she considers reasonable only that of his withdrawal into a monastery or to some other secluded location. Totally groundless are the fantasies which surfaced much later of a kidnapping by foreigners (at that time politicians had no hint of the importance of nuclear physics), or of a possible escape to Germany, the USSR, or another country to collaborate in research (at the end of the war, specific testimony would have been provided by his physicist colleagues. To illustrate the weakness of this hypothesis, we include a letter toward the end of this volume, *T/10,* written to us by Pontecorvo. Moreover, in a case such as this, Ettore could have discussed it with his family and there would have been no need to engineer the painful contradictions contained in his final letters).

It remains to be ascertained whether Ettore, once he decided to continue his project, returned to Naples or stayed (at least temporarily) in Palermo. Supporting the idea of his return to Naples are the reports that we have already examined: from the nurse, and from the cloistered convent of San Pasquale in Portici (the one from the Convent of Gesù Nuovo is not conclusive, as the Superior does not remember if Ettore arrived before or after the Naples–Palermo trip). The nurse could be credible, because she knew Majorana well. In the letter dated 22nd January 1938 from Naples, Ettore told his mother: "... I'm still in the [*hotel*] Terminus, but I will soon be in a boarding house... I have some good addresses of boarding houses which were given to me by the nurse." And Ettore could have gone to one of these boarding houses, avoiding detection through police examination (if they did this) of the hotel registers.

But puzzling is the testimony of Vittorio Strazzeri, the professor of geometry at the University of Palermo, who apparently traveled back from Palermo to Naples in the same passenger compartment with Ettore. He did not exchange words with the person who he believed to be Majorana, but with the third passenger — a man named *Carlo Price*, according to Tirrenia's records. He was not an Englishman, but some type of shopkeeper, or of lower status, who spoke Italian "like we from the South do". So one might suggest that Ettore did not leave with that ship, but may have sold (or given) his ticket to another passenger, who we will call "Mr. X". Sciascia, in fact, put forth the interesting hypothesis that Mr. X was mistaken by Prof. Strazzeri for Carlo Price, while the latter (English and quiet) was mistaken for Majorana.

Likewise, it appears that Tirrenia never produced Ettore's return ticket, which they had declared to have found. The weave of fleeting light and ambiguous shadows created by Majorana had already taken on a life of its own. Now the *facts* themselves, subsequent to his disappearance, carried forward Ettore's "project", and we shall soon see the most perplexing examples of this.

Around 1969, rumors reached us on several occasions that Ettore had remained in Sicily following his trip from Naples to Palermo, and of a subsequent emigration to Argentina. But because we did not investigate these rumors, we could not put faith in them. As Emilio Segré wrote to us in January 1980, from Berkeley: "As to those who have seen Majorana in various places: There are many people who have seen the Dauphin son of Louis XVI, the relatives of Tsar Nicholas, etc. The phenomenon is not uncommon…" Indeed, as you can imagine, much of the "testimonies" are either extravagant or simple-minded. We went on receiving pieces of information about Majorana's fate also during the last 20 years, when we had few time left for further serious checks (or, we could say, few new piece of information demanded more than scarce attention to us): in any case, we do not report them.

We may quote an old one, because it was given to us in good faith, and because others have given it a certain amount of attention. Several years ago, a respected professional put us in contact with the secretary of a school in Catania, Antonio Versaci, who around the years 1950–1951 was the barber for the Jesuit priests of Villa San Severo, performing weekly visits to cut the priest's hair. One day he encountered a priest that he did not know. The priest was not tall, was on his way to attend spiritual exercises and was boasted to him as being a professor of physics "with many patents."

Intrigued, Mr. Versaci convinced him to get his hair cut, and seized the opportunity to ask him questions about atomic energy. The priest confided that he was, in fact, a colleague of Fermi and Segré, and explained to him several concepts of science. The conversation ended with a joke. When Versaci thanked him for the clarity of his explanations, the priest replied: "Didn't you know that you address professors as *Chiarissimo*?[3] So we *have* to be clear!" This witticism, in a moment of leisure with the barber, could have come from Ettore. But the priest's appearance, as described by the witness (white hair, round face), did not fit Majorana, who would at that time have been around 45 years old. Also, the witness claimed to have "recognized" him, but on the basis of which photographs? How long had it been since he had seen them? It would have been odd for Ettore to choose Catania for his spiritual exercises. At last, it did *not* turn out to us that Majorana devoted his life to religion, even if later on we received a number of other reports (unfortunately not easily verifiable) to this effect.

Possible Motives

But let's return to what factors might have contributed to Ettore's decision; or at least to those usually promoted.

Amaldi holds that the reasons were personal; connected, perhaps, to the difficulty of finding a companion

[3] A respectful manner of addressing university professors, which also means "Very clear" in Italian.

compatible with his way of life. Recalling the gastritis that Ettore suffered from, the first symptoms of which had occurred during his period in Leipzig, Amaldi wrote: "The origin of this illness is not clear, but family doctors linked it with the principles of nervous breakdown."

Piero Caldirola believed, however, that his "stomach pain" concealed a serious illness at the physiological level, not neurological.

What we can say, on the basis of the letters we have in our possession, is that during his stay in Leipzig and Copenhagen, a few changes began to appear which could be revealing. For example, starting with the letter of 18th March 1933, he stopped signing his name with *aff.mo (your loving) Ettore*, as he had always done since he was a child, and began signing with only his name, and in a definitive manner — except for his *last* letters.

He also changed his opinions, as previously noted, on the subjects of fascism and Nazism. This was likely due to the effect it had on his ability to live alone (and probably well) in the well-organized, comfortable "courteous and friendly" German city of Leipzig. The news that he sent home regarding the political events of the time seem mainly inspired by a determination not to alarm his family, but it is clear that his initial judgment (on the catastrophic domestic situation and the indifference of the people) then changed. This is evident even though the historical justifications he relied upon to explain the events around him do not contain any hint of enthusiasm, as noted by Sciascia. A fan of psychoanalysis might even conclude that the rise of a leader (like Hitler) was a welcome reaffirmation of the "father figure" for Ettore.

At a certain point, the letters to his mother began to include more economic, statistical, and political commentary. His letters were being read by everyone, and certainly in Ettore's house there were attentive and valuable readers, if not interlocutors. But it seems equally true that Ettore increasingly wished to impress his own style in his letters, soon freeing himself (while still remaining affectionate and sincere) of the themes of laundry and health.

We recall that Majorana made a final attempt to accept a "normal" life by participating in the Competition for the professorship at the end of 1937. His intention to reintegrate into society seems supported also by a small sign: in his letters to Gentile (starting in 1929) and his uncle Quirino (starting in 1931), Ettore had never included the "fascist era" year[4] following the date. But in those from November 1937, the month of his appointment as professor, there the "XVI" pops out. Setting aside these considerations, Ettore may have been shaken by the suicide in 1933 of Paul Ehrenfest, the great Dutch physicist and thinker we have already

[4] During Fascism in Italy it was common to denote the number of years since the March on Rome in 1922 using Roman numerals.

mentioned and with whom Ettore had become friends in Leipzig. He was certainly deeply affected by the death of his father in 1934. However, we do not, believe that his concern for the legal misfortune of his uncles Giuseppe and Dante was a determining factor, or the source of Ettore's distress (Dante, along with his wife, was imprisoned for three years before being acquitted in 1932). That was the famous "crime of the cradle" that took place in 1924, when Ettore was only 17 and on which Sciascia, for our purposes, already had the last word. We add only that Ettore was *very* worried. Gleb Wataghin, the accomplished Italian physicist of Ukrainian origin and founder of Brazilian physics, left evidence of this in 1975 at the University of Campinas, S.P., in an interview conducted by the Institute of Physics that bears his name. His speech (in Portuguese) is colloquial: "Ele me contou, nós éramos bastante amigos, que tinha un processo muito sério relativo ao seu tio – ele gostava muito do tio – que foi acusado de uma vinganza…, onde resultou que um menino [Cicciuzzu Amato] foi quemado, por negligência da dona de serviço…Ele então disse: *Eu não acredito nos advogados, são todos estúpidos; vou eu escrever a defesa do meu tio: eu conheço a coisa, conversei com ele.* Estava escrevendo…"[5] Above all, Wataghin's recollection is inaccurate

[5] "He told me, since we were fairly good friends, that there was a very serious court case against his uncle — he was very attached to his uncle — who had been accused of a vendetta…, in which it appeared that a child [Cicciuzzu Amato] had been burned to death due to negligence of the

regarding the time, as he seems to refer to 1933. But by 1932, the trial had already concluded. Ettore's involvement, while sincere and undoubtedly intelligent, was unrealistic. The distinguished lawyers would almost certainly have paid little attention to an "amateur" who by the end of the case was still only 25 years old.

Still, let us profit from Wataghin's recollections and return to the atmosphere in Leipzig: "In Leipzig, where Heisenberg worked, I met Jordan, Debye, Max Born who was just arriving there, and Ettore Majorana, a young man who *pareceu, como era realmente, un verdadeiro gênio...* The camaraderie, the friendship that exists between scientists... manifested itself, for example, in the way our scientific discussions took place — similar to sporting events. In Leipzig, we would gather for two-hour seminars from two to four in the afternoon. In the morning, theoreticians sleep.[6] Afterward, we would play Ping-Pong in the library, on a table meant for students. I can say that the champion was Heisenberg. Then we would go to a pub, and perhaps play chess. We also played chess at the Institute of Physics. Because Heisenberg was one of the directors, nobody complained about our playing Ping-Pong or chess in the library, but this would

---

housekeeper... He then said: *I have no faith in lawyers. They are all stupid. I will prepare the defense for my uncle. I know the situation, I've spoken to him.* He was writing..."

[6] For those who wish to know what physics is, we recall Orear's definition: "Physics is what physicists do late at night."

have been unthinkable at that time in other institutes... The seminars were attended by people from all over the world. For example, I remember that once during a seminar held by Norzig and one of his colleagues, they were forced into a very challenging discussion *depois das perguntas que faziam o Heisenberg e o Ettore Majorana*[7] [following questions asked by H. and E.M.]." Again, declares Wataghin in the interview, "I remember Majorana in particular who, according to the judgment of many and in particular Fermi, was an exceptional genius... He was ill, suffered from an ulcer, and consumed almost exclusively milk; he did not participate in sports or gymnastics; he often took long walks by himself. He was not very communicative. But every now and then we would meet him on Saturdays. He was very critical. He found that *toda gente que ele encontrava era não preparada, ou estúpida, ecc* [everyone he encountered was ill-prepared, or stupid, etc.]. He spent a lot of time on statistical laws as they applied to nuclear matter... The exchange symmetry between protons and neutrons might be complete, including charge and spin, or regard only the charge or the spin. This had not been proposed or studied by others. And the exchange symmetry of the sole positions

[7] Pontecorvo also remembers: "Majorana was a pessimist by nature and eternally dissatisfied with himself (but not only with himself). During seminars he usually remained silent. He sometimes broke his silence to utter a sarcastic comment, a paradoxical, though relevant, observation. I remember that in seminars he often terrorized famous foreign physicists..."

of protons and neutrons (without touching the spin) allowed the statistical comprehension of why nuclear material should have a constant density... This meant that Majorana's theory held a great advantage over that proposed by Heisenberg."

Giuseppe Occhialini Speaks

New, poignant light on the thoughts and feelings in Ettore's heart and mind at the beginning of 1938 comes to us from Giuseppe Occhialini, another celebrity of Italian physics (*experimental*, this time) who, for his role in the discovery of the anti-electron first, and the π meson later, would have merited *two* Nobel Prizes. Among the physicists Ettore knew, Occhialini was perhaps the one he found most congenial because of his original thinking and intolerance of conformism, convention or "normalcy"; perhaps also because from the age of 18, he had been fond of saying that a life lived as such was best to put to an early end. And — as an event probably unique — Ettore opened up to Occhialini in January 1938, albeit through allusions and using a jargon comprehensible only to them. Indeed, Occhialini, returning by ship from Brazil, took advantage of a stop in Naples to quickly visit Carrelli's Institute. Here, he met Majorana for the first (and last) time. "It is curious," Occhialini said, "that I meet you now in Naples, on my way back from San Paolo, when we were neighbors for many years while you were in Rome and I was in Florence, and we never saw each other." "You've arrived just in time to meet me," answered Ettore, tersely, "If you had arrived any later you would not have found me. Because *there are those who talk about it, and there are those who do it.*" Ettore was cryptic, but not really. And certainly not for Occhialini, who immediately felt a hint of reprimand for his chattiness in the face of problems concerning life and death, survival and suicide. Occhialini understood, and was moved by the trust that Ettore had shown him, but he had to leave. He reflected on this phrase all the next night, but in vain; without having been able to help Majorana.

It is remarkable that Ettore confided in Occhialini. Almost equally extraordinary, however, is that Occhialini later spoke about it,[8] although it was only after more than 50 years, in 1990. Ettore's confession to Occhialini is the element among those known to us which most strongly points in the direction of a suicide, but this too must be treated with caution. In all likelihood, Ettore wavered between suicide and a mere *disappearance.* And in expressing himself around colleagues, Ettore often preferred to use the most succinct terms: those that would leave them the most surprised or astonished.

---

[8] The person who deserves the credit for obtaining this hard-won trust from Occhialini is Bruno Russo.

8

Did Majorana Seek Refuge in Argentina? — Some Documents

I became "one". I myself.
I who wanted to be this way.
I who felt this way.
Finally! …
No longer "Gengè". Someone else.
This is exactly what I wanted.
LUIGI PIRANDELLO
(One, no one, and one-hundred thousand, 1926)

From the World of Science (The First Testimony)

And now, finally, we will reveal that Ettore went "to America or elsewhere". Or rather, to Argentina; in fact, to Buenos Aires.

Maybe.

Some witnesses say so, and these appear to be the most credible we have. This will be for the reader to judge. In this chapter, we will once again not only encounter grand personalities and Nobel Prize laureates, but also letters which are lost, testimonies given and then withdrawn, and long silences. As stated earlier, Majorana's *project*, once underway, took on a life of its own and continued to be fueled by the very events. And with the same lights and shadows Ettore had desired from the beginning, with the same ambiguities. Almost by the same Direction.

The disappearance of Ettore Majorana has provoked periodic waves of interest from the media throughout the decades. With each wave, the excitement among various media would grow, swelling increasingly until it saturated the public's attention, and then fade out. One report, which follows one such wave, no longer has any chance of gaining traction — even if true or groundbreaking.

Following a flood of articles in 1975–1976, a *new* piece of information appeared in the press. But it arrived too late, in October 1978, and it was fortunate that no one paid much attention to the "sensational" news that emerged after an avalanche of fabrications, repetitive articles, and trivial comments on the subject of Ettore's disappearance.

We say *fortunate,* as Ettore's family said; and perhaps Ettore would have said.

Sometimes there is irony in fate. Did fate itself collaborate in the *project*?

In any event, on 8th October 1978, Gino Gullace, a correspondent from New York, recounted the following story to the weekly magazine *Oggi* (which would be released on the 14th): a well-known Italian physicist, a friend of his, had spoken to him in April about a Chilean colleague who claimed to have heard of the presence of Ettore Majorana at an Argentinian restaurant in Buenos Aires. He continued:

> "What was the name of the Chilean physicist? I was given the name Igor Saavedra. I phoned him in Chile, but he knew nothing about it... There must have been a mistake... In fact, several months later an Israeli scientist [*we'll call him Mr. Y*] explained to me that the physicist who had made the revelation was not Saavedra, but a professor named Carlos Rivera.[1] Rivera was the head of the Institute of Physics at the Catholic University of Santiago, was 52 years old, and had studied physics in Germany where, for 22 years, he'd worked alongside one of the greats of modern physics, Werner Heisenberg. Here is the story that Rivera told me: *"In 1950 I went to Buenos Aires with my wife and lodged at the guest house of a woman[2] named Frances Talbert. This woman had a son called Tullio Magliotti who had a degree in electrical engineering. The day before I left for Germany, I was in my room and writing on a few sheets of paper. I was working on Majorana's statistical laws, and his name was written in large letters on one of the pages. Mrs. Talbert, seeing the name, exclaimed: 'Majorana? But this is the name of a famous Italian physicist who is a good friend of my son. In fact, they see each other often. My son told me that he no longer dealt with physics, but with engineering'. The conversation continued for a bit longer*

[1] Appears erroneously as "Ribera" in the text.

[2] A friend of Rivera's parents.

and the woman added, 'Majorana told my son that he left Italy because he did not like Enrico Fermi. In fact, he said more: that he did not even want to hear Fermi's name. This aversion, according to Eng. Magliotti, stemmed in part from the fact that Fermi was a 'difficult type', and partly from the fact that he had played an important role in building the atomic bomb. The conversation was interrupted by a phone call from her son. Perhaps he did not like that his mother had spoken to me about Majorana, and that I wanted to meet him. Mrs. Talbert, in fact, did not return to resume the conversation, and because I was to leave for Germany the next day, I could not meet her son nor continue the discussion with her. Four years later, in 1954, I went back to Buenos Aires and again went to see Mrs. Talbert. But the door to her house was nailed shut, and no one was there. I asked the neighbors for news of them, and they told me that the mother and the son had suddenly and mysteriously disappeared. Mrs. Talbert had been openly anti-Perón; I believe that she and her son may have been eliminated by Peron's police. I then went to check the register of engineers, but Magliotti's name was not on it. Murdered? Did they escape to some deserted part of Argentina? I don't know." Professor Rivera's story does not end with Mrs. Talbert and her son. It has a sequel: "*In 1960,*" he continued, "*I returned to Buenos Aires a third time, lodging at the Hotel Continental. This is where the napkin incident occurred. While I was seated at the table writing formulas on one of the napkins, a waiter said to me: 'I know another man with an penchant for writing formulas on paper napkins, as you do. He is a client who comes in occasionally to eat or to have coffee, and is called Ettore Majorana. This man*

> *was a very important physicist and fled from Italy many years ago.' This second episode, though less important than the first, convinced me that Majorana had to be in Argentina. The waiter did not know where I might find him."* Rivera did not know the name of the waiter. "*But*" he adds "*Perhaps by going to Argentina his trail might be found.*"

Actually, the beginning of Gullace's article contained several inaccurate or misleading statements that we have spared the reader. There are, however, still a few things to examine and others to verify. Some of the inconsistencies are apparent — for example: (a) the reason for Ettore's escape from Italy could not have been an alleged dislike of Fermi. If that had been the case, Ettore could have returned after just a few months, as Fermi left Italy permanently at the end of 1938 to go to Stockholm for receiving the Nobel Prize. Also, Majorana's departure amounted to a complete break with everyone, including his family; (b) in March 1938, Ettore could not have known that Fermi would, several years later, play a role in the development of the atomic bomb.

The person who informed us of the article's imminent release in the magazine *Oggi* was Remo Ruffini. The following day, 9th October 1978, we wrote directly to Carlos Rivera. On the 18th of that month, from Pontificia Universidad Catolica de Chile, we quickly received a reply (in Spanish, this time):

"Santiago, Octubro 18 de 1978 — Estimado Dr. Recami — Hoy recibí su amable carta en la que me solicita si lo afirmado por el periodista Mr. Gullace en la Revista "Oggi", no. 41, Oct. 14, 1978, pagg. 95–97, corresponden efectivamente a lo afirmado por teléfono por mí.

Puedo dicirle que lo afirmado por el Sr. Gino Gullace corresponde al conoscimiento que yo tengo del futuro de Ettore Majorana.

No tengo mayor información que la que le indiqué al Sr. Gino Gullace.

Puedo sí asegurarle que la señora Talbert vivía aterrorizada en su departameto en Buenos Aires, debido al regimen opresivo de Perón. — Cualquier otra noticia que pueda obtener posteriormente se la comunicaré a la brevedad posible.

— Lo saluda muy atentamente.

Carlos Rivera Cruchaga,
Director, Instituto de Física."[3]

So Rivera confirmed the information from the journalist. But is he trustworthy? Who passed this

[3] *...Today I received your kind letter in which you ask if what the journalist Mr. Gullace wrote in the magazine "Oggi", edition 41, of* 14th *October 1978, p. 95–97, corresponds to what I told him by phone. I can tell you that what Mr. Gino Gullace wrote indeed corresponds to what I know of the fate of Ettore Majorana. I have no further information beyond that given to Mr. Gino Gullace. I can assure you that Mrs. Talbert lived in terror in her apartment in Buenos Aires due to the oppressive regime of Perón. If any other information comes into my possession in the future, you will be notified immediately. With many kind regards, — Carlos Rivera C.*

information to Mr. Gullace in New York? And who is the Israeli scientist who we have provisionally called Mr. Y? Let's proceed systematically, as we said, but premising that we also wrote to the Hotel Continental in Buenos Aires, and received no reply.

A Verification: Tullio Regge

Remo Ruffini had informed us of the role of Regge, who had had the opportunity to go to Chile and get an impression of Rivera. Tullio Regge was a well-known Italian relativist, and a leading theoretical physicist. On 28th November from Turin, on the letterhead of the Academy of Lincei, he pointed us toward an important testimony — the one we needed. Here it is:

> "Dear Recami — I answer you late because you wrote to Princeton, rather than to Turin where I have returned.
>
> The colleague from Tel Aviv "are" two. One is Yuval Neeman, who gave me an inexact version but one that gained my attention. Y.N. had heard it at a party in Texas hosted by Wheeler.[4] W learned of it in Varenna (I believe) from Robert — now Yehuda — Meinhardt, who is now Israeli, but was previously Chilean. R. Ruffini knows him also. R.Y.M. told me personally, while I was in Tel Aviv, that Carlos Rivera, a professor at

---

[4] John Archibald Wheeler, one of the most fascinating characters in physics, and to whom J. Klauder dedicated his book *Magic without magic*.

the Catholic university in Chile, had an experience in Argentina similar to that which later appeared in "Oggi." By chance, shortly afterward I went to Chile and I made a point of speaking with Rivera, who is truly the key person in the whole affair. Others have reported stories which are, more or less, conspicuously inexact based on Rivera's accounts, which were subsequently published in "Oggi". Published, I would say, without alteration. Rivera confirmed the meeting in Buenos Aires with Tullio Magliotti's mother (Mrs. Talbert), during which she'd stated that her son was a friend of a certain physicist called Majorana, who wrote formulas and quite disliked Fermi (reciting from memory) and who left Italy for this reason. What is strange is that Rivera also said he had met a chef at the Hotel (Continental?) who more or less told him a similar story. What can I say?

Rivera certainly did not seem like a pathological liar. He is a respected professor at Catholic, educated at Gottingen, of high culture and does not seem the type to tell lies. I'd recommend that Miss Majorana contact C. Rivera directly at *Catolica* in Santiago, Departamento de Física... I don't know what else. I was struck by Rivera's obvious hostility toward Fermi, and he attributes hostility toward F. to the presumed Majorana. Rivera also has a theory.

Rivera's theory

1) E. Majorana really was in Buenos Aires at that time.
2) It is a fact that Magliotti was anti-Perón and that many people disappeared, having been kidnapped and

> murdered by Perón's police. In his opinion, Magliotti and his mother were victims of this. There is no longer any trace, according to Rivera, of the two, T.M. was an engineer.
> 3) E. Majorana would have also been involved in T.M.'s political affairs and would have (?) suffered the same fate.
>
> I've reported these things just for completeness, and do not assume any responsibility for them. It would be better for you to speak with Rivera and/or hire someone to quietly conduct a series of investigations in Buenos Aires. — Best regards, T. Regge."

We had already written to Rivera. We had also contacted Y.R. Meinhardt, writing to him at the University of Tel Aviv, but the letter was returned ("Inconnu" unknown recipient — secretaries are not always efficient).

Point 3 of Rivera's theory doesn't hold water. Rivera himself stated that he had met the *chef*, or waiter, in 1960. And if a waiter remembers seeing Majorana five years after the end of the Perón dictatorship (which lasted from 1946 to 1955), it means that Ettore *did not* suffer the same fate that the Talbert family is feared to have met — the Talberts disappeared in 1954, when the regime was near its collapse.

Regarding "quietly" conducted investigations, we immediately contacted the director of the Department of Physics at the University of Buenos Aires, Giulio Gratton. Giulio is one of the many scientist-sons of Italian astrophysicist Livio Gratton, who had an important role in the development of this science in Argentina. The investigation would require some time, and the answer came after six months.

The response was negative; it would exclude even the existence of a Hotel Continental, one of the oldest and most famous hotels in Buenos Aires! A mistake? Perhaps, but one influenced by the legitimate desire, then still common in Argentina, to distance oneself from any search for a person, as it could dangerously appear like a search for a *missing* person.

The same day that Regge sent us his testimony, Carlos Rivera again picked up his pen to write — this time to Maria:

> "*Santiago, November 28th, 1978 — Muy estimada y respetada Sra. María Mayorana: He recibido su carta* ....[5] I received your letter. I ask sincerely that you forgive me for the delay... I absolutely do not want to be put back in contact with the journalist Gullace, nor will I give any public statement. Sadly, with so much time having passed, I do not remember further details about Mrs. Talbert, neither her appearance nor the face of the waiter at the Hotel Continental in Buenos Aires. In my opinion, the terror

[5] For the original text, see Part two, Chapter 12, Letter *T/A4*.

that Mrs. Talbert displayed was due to the horrible persecution that she had been subject to under Perón, and I am almost certain that her son died a victim of this persecution. It is not improbable that Dr. Ettore Majorana escaped the Perón persecution, and personally I do not know his subsequent fate (did he return to Italy?). *[When writing to Ettore's sister, perhaps out of sensitivity, Rivera abandons point 3 of the theory to which Regge referred in his letter. And we can only agree. Again, perhaps out of sensitivity, Rivera then went on to provide a* different *theory]* ... — Please forgive me if I cannot ease your anxiety, since for these many years passed I cannot provide you more details. But a doubt has come to me; that some wretched person could have passed himself off as the great physicist Ettore Majorana, pretending to be your brother in order to profit from his immense prestige... These things happen occasionally in history when the true circumstances of a disappearance are unknown [*While he writes, Rivera seems to convince himself more and more*]. The hypothesis that the name was false — *he adds* — is probably the most plausible. Mrs. Talbert was of advanced age (she was a friend of my mother, who had met her on a trip from France to Argentina); she was quite sure it was Ettore Majorana, of whom she was quite fond. Despite the fact that she had never met him when he worked with Fermi in Italy, she had so much confidence that everything seemed certain to her. For any other information that I receive which might be useful, you can count on me that I will not pass any information to the press. And if I had known that Ettore had a sister, I would have written directly to

> you, without rendering any of it public domain. — *Con muchísimo afecto la saluda muy atentamente*, — Carlos Rivera Cruchaga, Director etc."

Carlos Rivera's new theory may convince him, but it does not convince us. Majorana (which Rivera's typist writes as *Mayorana*) was then known only to specialists. A compulsive liar would have provided a name much better known to the public. A cheater, assuming that he was aware of the Majorana case and that he could fool an engineer like Magliotti, would have attempted to establish himself within the scientific community. The name of a complete unknown would have served little purpose with the ordinary citizen. Instead, a few other accounts (such as that of Italian theoretical physicist Daniele Amati, who worked in those years in Buenos Aires) assure us that if Majorana had really lived near enough to Buenos Aires to have regular opportunity to frequent a hotel in the city, he carefully avoided any contact with the physics community. According to the same Rivera, Ettore dedicated himself to engineering, like his friend Magliotti.

The General's Intervention

The track number one was therefore coming to a standstill — like the search so many years before in Naples.

Ettore would have been 72 years old in 1978, and could well have still been alive. Perhaps it was this thought that rendered us, unconsciously, a bit lazy in our investigation. Who would have had the courage to submit Majorana to the violent clamor that his discovery would generate?

But on 31st July 1980, we received a telegram: "Dear professor Recami Stop Tullio Regge has told me that you have a written statement of the facts uncovered by him. I was involved indirectly in the investigation of Majorana's disappearance and speculative emergence in Argentina Stop Could I have a copy Stop Best regards — Yuval Neeman."

Those familiar with the sender will understand why the telegram gave us a jolt. The Israeli physicist Ne'eman, prior to becoming a scientist, had been a career general in the military. Only in his later years did he decide to devote himself to theoretical physics, becoming within just a few years capable of making first-rate contributions to the field. He maintained his old ties, however, and would go on to assume positions of great responsibility in the Israeli government and intelligence services. The involvement of Neeman, therefore, could be decisive. Certainly no one could have hoped for more.

We responded, addressing our letter to Israel. On 20th October 1980, he wrote to us from the University of Texas (in Austin, where we had first met him):

"Dear Erasmo: — Your Majorana paper and the answers to my queries to you have all caught up with me.

My interest in Majorana was first awakened by conversations with the late Racah, who told me about the tragedy. In 1975 I learned about the renewed interest in Italy after Shasha's [Sciascia] novel (I know this is not the correct spelling but I don't have it before me, so I write the name phonetically).

I also read Amaldi's articles answering Shasha [Sciascia], etc. I am responsible for the revival of the Argentinian version. Wheeler first heard it from Meinhardt (a Chilean Jewish Physicist who had meanwhile migrated to Israel) at Varenna[6] *in 1977, but confused Carlos Rivera*[7] *with Saavedra. Realizing the importance and freshness of the issue in Italy, I related it to Regge, who had somebody call up Saavedra in Santiago. Saavedra said it wasn't him. I looked up Meinhardt in Israel and had him meet Tullio [Regge] when Tullio visited me there in May, 1978. Tullio went to Chile from Tel Aviv and got the details from Carlos Rivera. When I wrote to you, I did it after Tullio told me he had reported it all to you, and I wanted to have the precise results of his inquires. I would certainly like to help check on these facts, but it seems very hard. Argentine has had so many upheavals! However, I shall try and find out whether the Jewish family, mentioned by Rivera as having been in contact with Majorana at the time, hasn't ended up in Israel. If I do discover something, I shall certainly let you know (and Miss Majorana) immediately. With kind regards, Yours cordially, — Yuval."*

---

[6] Varenna on Como Lake: a center (like Erice) for International conferences.

[7] Appears erroneously as "Ribera" in the text.

In the meantime, we confided to him that we had received a few other independent reports of Ettore's possible presence in Buenos Aires. And Ne'eman was quick to confirm his availability (writing again from Tel Aviv: "November 23rd, 1980 — Dear Erasmo, To enable me to pursue the matter of the "Argentinian" version of the Majorana case, could you send me a copy of the relevant documentation you mention in your letter of 19th September 1980, including Tullio Regge's report regarding his conversation with Carlo Rivera? Best regards, Cordially yours — Yuval"). Although it took us more than two months to do so, we sent him all the material requested. But nothing came of it — track number one had truly gone cold.

The investigation, however, had already yielded another, independent lead.

From the World of Art (The Second Testimony)

Taormina, 1974. Some say that Taormina no longer attracts the "high society", as it once did. But in the summer it is still a preferred destination for some of the biggest names in culture — Italian, or otherwise.

Ms. Blanca de Mora, widow of Asturias, the Guatemalan writer and winner of the Nobel Prize for Literature in 1967, batted her eyes at celebrity friends when they mentioned the story of Majorana: "But why do you pose the question of Ettore Majorana? In Buenos Aires there were many of us who knew him. While I lived there I would meet him occasionally at the home of the Manzoni sisters, who were descendants of the great novelist." Ms. Blanca Asturias had lived in Buenos Aires up until the early 1960s. The sisters Eleonora (a mathematician) and Lilò (a writer), probably *not* descendants of Alessandro Manzoni, had held a cultural discussion group there. Ettore was a friend of Eleonora, the mathematician. But Asturias' widow had lost touch with all of her acquaintances after leaving Argentina for France (some people, in her view, had even disappeared in recent years).

This we learned only in 1980, from painter Carla Tolomeo. Maria Majorana then wrote to Blanca de Mora Asturias in Paris. For six months, there was only silence. Then Mrs. Asturias replied that yes, she remembered the Cometta-Manzoni sisters well, but not Prof. Majorana. She washed her hands of it, in fact, and suggested that Maria write to her sister, Lila de Mora, who resided in Caracas and who may have a better memory of it. However, mutual friends recommended caution, confiding to us that the widow feared that a publicized search for someone (Ettore) by Maria may bring harm upon those who had remained in Buenos Aires.

Ms. Lila de Mora's address, on the other hand, could be invaluable. Maria immediately sent it to us from Rome, even sending the original letter that she had received from Paris. Imprudent? In Italy, the mail is often slow, but letters are not lost. This one was lost.

Some say that fate is blind. We do not think so. Initially, fate played on irony's edge, but now was mocking. Fate intervened, indeed, quite heavily.

After, a few more years passed like this; how could we not surrender to some degree? Einstein, observing the difficulties that nature places before those who seek the truth, reassures us: "God is subtle, but he is not malicious." However, we must admit that, as regards our small case, a certain prayer, a special recommendation, had found an audience in the heavens.

In Latin America

In international interactions among scientists, academic rituals (which in the field of physics can be quite informal) are occasionally interrupted by more mundane events. The Consulate of Italy might send invitations for a conference on, say, E. Majorana. In 1985, this is what happened in São Paulo, Brazil.

Later, this topic is discussed again among physicists at the university, this time with no reservations. We were in Latin America, and an Argentinian scholar, Professor Foglio, smiled at hearing the name of Eleonora Manzoni; she was a student who had attended his lectures in mathematical analysis. An Italian colleague, in particular, decided to get actively involved. Her name was Maristella, a professor at the University of Campinas, and daughter of the Italian astronomer Fracastoro.

The first thing to be done was obvious: write again to Ms. Asturias. The letter was sent at the end of March. For once, fate moved in our favor. A reply was received immediately afterward. It was sincere, written without prior consultation with distant friends and, dispensing with her former misgivings, was finally positive:

> "Paris, le 7 Avril 1985 — Mme Maristella Fracastoro, Institut de Physique, UNICAMP — *Chère Madame: Il y a déjà quelques années j'avais reçu une lettre, sur le cas du Professeur Ettore Majorana, et de son amitié avec mes amies Cometta Manzoni.*
>
> *Je vous ai répondu que mon amie Eleonora Cometta-Manzoni était certainement amie de Majorana, mais qu'elle était décedée déjà depuis quelques années. Eleonora et sa soeur habitaient à l'époque à la Rue Santa Fé 2189, Buenos Aires, Rép. Argentine. Peut-être on vous donnera là leur nouvelle adresse. L'autre soeur mariée à un ingénieur Vénézuélien habite Caracas, Venezuela: Loló* [Liló] *Cometta-Manzoni de Herrera; elle est professeur des Lettres à l'Université de Caracas, mais je ne retrouve pas son adresse car je connais sa maison, sans m'inquiéter de l'adresse. Mais il ne vous sera pas difficile de la rencontrer car ce sont des personnes très importantes là bas. Ou, en dernière instance, adressez-vous à ma soeur Lila de Mora y Araujo de Gándara Casares, avenida... Buenos Aires; elle est liée d'une grande amitié avec ces soeurs ainsi que moi-même: mais j'ignore leur nouvelle adresse.*
>
> *Je regrette Madame Fracastoro ne pas pouvoir vous être plus rapide et directement utile. Je suis moi-même éloignée de mon pays où j'espère pourtant rétourner quand la succession de mon mari Miguel Angel Asturias, Prix Nobel de Littérature, Grand*

Officier de la Légion d'Honneur de France, Grand-Croix de Bolivar en Colombie et Grand Prix de la Paix, sera finie.

Vous souhaitant une bonne fin à vos recherches, dites à Mme Mayorana que le nom de son frère ne m'étais pas inconnu, mais nous avons quitté Buenos Aires en 1961: Eleonora Cometta, une profonde amie de ma vie et de mon coeur, était ancore vivante.

Je vous conseille d'écrire à ma soeur Lila de Gándara Casares.

Je me répète votre amie,

Blanca de Mora y Araujo de Asturias

P.S.: Au moment de fermer la lettre, j'ai la chance de retrouver l'adresse de la soeur de Eleonora Cometta-Manzoni (mathématicienne) Loló [Liló] Cometta-Manzoni de Herrera (Professeur de Lettres): calle…, Caracas, Venezuela."[8]

---

[8] "Dear Madam: several years ago I received a letter regarding the case of Professor Ettore Mayorana [sic] and his friendship with my friends Cometta-Manzoni. — I replied that my friend Eleonora C. Manzoni was certainly a friend of Mayorana, but that she had already been deceased for several years. Eleonora and her sister lived at the time in via Santa Fé 2189, Buenos Aires, Rep. Argentina. Maybe, there, they can give you their new address. The other sister, married to a Venezuelan engineer, lives in Caracas, Venezuela: Lilo' C. Manzoni de Herrera is a professor of Literature at the University of Caracas, but I don't have her address because I knew where her home was, without needing an address. But it will not be difficult to find her because they are very important people there. Or, as a last resort, you can speak to my sister Lila de Mora y Araujo de Gándara Casares, Avenida …, Buenos Aires; her friendship with these sisters was as close as mine, but I do not know their new address. — I am sorry, Ms. Fracastoro, that I am not able to be of more use. I, too, am far away from my country,

Finally some confirmation, and it was completely independent from that of Carlos Rivera! Let's keep this in mind because this second path will not be quick to yield others. This latter testimony, however — initially negative and then positive — has put us on alert. We shall therefore approach them with a more experienced eye.

Our Maristella Fracastoro now had two addresses she could use. She sent two letters in April, and the responses, both arriving in early July, are well thought-out and carefully edited.

Professor Liló, Eleonora's sister, replied from Venezuela: "Caracas, julio 4 de 1985 — De mi mayor consideración: Llegada hace unes días de Buenos Aires me encontré con su carta y una de la Sra María Majorana que ya he contestado. Desgraciadamente yo ni mis hermanas tenemos ningun conocimiento del Profesor Ettore Majorana, que segun ustedes era amigo de

---

where indeed I hope to return after the succession of the estate of my husband Miguel Angel Asturias, Nobel Prize for Literature, Grand Officer of the Legion of Honor of France, "Grande Croce di Bolivar" in Colombia and Grand Peace Price — In wishing her success in her search, tell Mrs. Mayorana (sic!) that her brother's name is not unknown to me, but we left Buenos Aires in 1961: Eleonora Cometta, a close lifelong friend who was dear to my heart, was still alive. — I advise you to write to my sister Lila de Gándara Casares. Renewing my friendship with you, — Blanca de Mora y Araujo de Asturias. P.S.: while closing this letter I was fortunate enough to find the address of the sister of Eleonora C. Manzoni (the mathematician), Lilo' C. Manzoni de Herrera (Professor of Literature): calle ..., Caracas, Venezuela."

nuestra hermana Eleonora. Lamento muchissimo por lo tanto no poder aportarle ningun dato con respecto a su paradero. De todas formas, si por alguna casualidad, investigando entre las amistades de mi hermana, llego a tener alguna noticia, puede usted estar segura que de imediato me comunicarei con usted. — Siento mucho no poder ayudar en su búsqueda y quedo a sus completas órdenes. — Liló Cometta Manzoni."[9]

Dr. Lila de Mora, Ms. Asturias' sister, answers from Argentina: "Buenos Aires, 10 de julio 1985. — Distinguida Señora: siento mucho no poder serle de utilidad en su delicado empeño en busca de huellas biográficas del profesor Majorana. Como usted bien lo dice, mi amistad con Eleonora Cometta Manzoni fué realmente la de dos hermanas, pero no conoci nunca al profesor. Es verdad que nuestras actividades eran muy diferentes. Yo era profesora de letras, pero asi y todo, fui muy amiga de otros compañeros de su facultad, como Rey Pastor, Enrique Butty, Juan Blaquiar, etc.; pero no conoci como le digo antes a Majorana. No obstante me puse en comunicación con las

[9] "With my greatest consideration: — Having arrived from Buenos Aires a few days ago, I found your letter and one from Mrs. Maria Majorana, to which I have already replied. Unfortunately, neither I nor my sisters know anything of Prof. Ettore Majorana, who you believe to have been a friend of our sister Eleonora. I am so very sorry not to be able to provide any information as to his fate. In any case if, while inquiring among the friends of my sister, I happen to come across any information, you can be sure that I will contact you immediately. — I am sorry I cannot help in your search and I am at your complete disposal. — Liló Cometta Manzoni."

hermanas de Eleonora, también muy amigas mias, pero tampoco ellas lo recordan. — Muy apenada, pues, por no poder serle de ninguna utilidad, me despido de usted con mi mayor consideración y de todos modos me pongo a sus gratas órdenes. — Cordialmente la saludo — Lila Mora y Araujo de Gándara."[10]

Reading this, it seems that Ms. Asturias was ashamed to inform her distant friend and sister that this time she'd had the "shallowness" of admitting to some truths; those that she remembered.

And it is quite clear that Mrs. Lila de Mora y Araujo de Gándara, before responding, had consulted with her friend Liló C. Manzoni during the visit by the latter to Buenos Aires (a visit Liló claims to have just returned from). Indeed, she admits it openly: *No obstante me puse en comunicación con las hermanas de Eleonora, también muy amigas mias...*

If the answer was a simple "yes" or "no", what was the point of seeking, or waiting for, international

---

[10] "Dear Madam: I am very sorry not to have been of service to you in the delicate task of searching for biographical traces of Professor Majorana. As you correctly state, my friendship with Eleonora C.M. was genuinely like that of two sisters; but I never knew the professor. It is true that our activities were very different. I was a professor of literature, but with all this I was good friends with the other colleagues of his faculty, such as Rey Pastor, Enrique Butty, Juan Blaquiar, etc... However, I did not know — as I mentioned earlier — Majorana. Nevertheless, I contacted Eleonora's sisters, also my close friends, but they do not remember him either. — So I am very sorry not to be of any help. With the greatest consideration, I remain at your service. — Sincerely yours. — Lila Mora y Araujo de Gándara."

consultation? Especially since Ms. Lila de Mora, a few years later in June 1990, would inform Bruno Russo — who went to Buenos Aires specifically to do further investigation — that yes, she had some memory of Majorana. She recalls the following episode: one evening Eleonora phoned, telling her "At this time I am unable to join you because Majorana has arrived". And, when asked about his physical appearance, she answered: "small, thin, reserved..."[11]

The letter from Ms. Liló Manzoni is not as open: *De todas formas, si por alguna casualidad, investigando entre las amistades de mi hermana, llego a tener alguna noticia...* But with this it seems that she wants to share the responsibility for what she writes with the friends of her sister. Perhaps she is also considering the possibility of taking a different approach in the future. But suddenly, as if to ward off further investigative letters, she adds: *...puede usted estar segura que [yo] de imediato me comunicarei con usted.*

From the World of Critics (The Third Testimony)

The Milanese critic Giancarlo Vigorelli, known for the help he provided to the "dissident" writers, both from countries like the USSR and from Latin America, occasionally met with Argentinian refugees. Together

[11] The image that emerges from the fog of her memory is, for us, strongly suggestive. But the problem is always the same: does she remember the exact name? And does the name really refer to "our" Majorana?

they would engage in discussions and conversations, and exchange news about mutual friends. Suddenly, *en passant*, some bits of information surfaced: "Ah, you know, that Ettore Majorana ..." This was in 1982.

This is the same way that the Critic mentioned it to us — in passing. And since then, although we intended to do it several times, we have not asked him anything more.

A genius, for the gifts he has, becomes a bit of a "public figure" and must concede to the interest others have in him: to the *legitimate* interest. Indeed, we have attempted to solve the "mystery of Majorana". With the consent of the Majorana family, we have rendered public all of the available documents. And perhaps we've moved close enough to the resolution. *Enough* meaning "as much as needed."

At the risk of being at the threshold of Ettore's personal life, it is best to tread the final steps of his path lightly: And let each of us use the steps of his own reason, intuition, conscience, and heart.

Why?

The human story of Ettore Majorana has been linked by some to a "rejection of nuclear weapons." As we have shown, there is no clear evidence to support this connection.

All of our research suggests that the Ettore's disappearance represented an escape from being a Pirandellian "puppet": an escape from his entire environment. Due to his exceedingly sensitive nature, he already suffers from a difficulty with human contact. His hypercritical exactitude and shyness further constrains him. He cannot bear the extra "external" weight resulting from a sense of duty to act out the roles expected of him — professor, physicist, and son. Thus Ettore, searching for inner salvation and equilibrium, feels the need to detach completely from every aspect of his being. Sometimes salvation lies in escape. An escape that would unavoidably require an irreversible break from all relationships, even from his family, and even from mainstream science.

And, if it is true that he declared that no longer wished to hear of Fermi, his reaction to "the Pope" would have represented nothing more than a camouflage to conceal a much deeper rebellion against all domineering *figures*; against *all* the rules, the implicit impositions, the constraints. We must not exclude, within the family, the mother figure — of his mother, however, Ettore would never have spoken ill.

In the end, we should not forget that Ettore's personal problems must have been so powerful and unyielding as to render meaningless any prospect of winning Nobel Prizes, or of keeping the solace of some dear family affections.

"He remained for his entire life," said Angelo Majorana of his cousin Ettore, "a prisoner of lucid rationality

> and cold calculation... But his devaluation of the sentimental world was only in appearance and forced. Behind that convinced insignificance of "doing", his detachment from writing, from speaking, from leaving behind any traces of himself that could be attributed to him, he was hiding something deeper and intimately more dramatic: the sentiment of the (dangerous) insufficiency and partiality of the *logos*..."

Recent accounts, possibly credible but not investigated by us, claim that Ettore was in Cilento following his disappearance, or that he died at an advanced age at a convent near Viareggio or Pisa. Some have even described to us how Majorana abandoned the Tirrenia steamship from Naples to Palermo before its departure; how he went to Germany, then left for Argentina, and subsequently returned to Italy — perhaps indeed to enter a convent (things of which we have no proof, but that we cannot exclude. He, who is able to glimpse into the exciting perfection of nature, observe its sublime beauty and the depth of its laws, may have ultimately sought refuge in the Supreme Mind Who revealed such a nature to him.)

Nonetheless, for most the drama of Ettore's disappearance is, at present, associated with the idea of rejection, of "escape" as a response to the danger of certain technological applications of scientific understanding. Let us say, then, that here we have a problem as old as the world itself. In 1903, Pierre Curie concluded his speech at his awarding of the Nobel Prize with these words: "It can even be thought that radium could become very dangerous in criminal hands, and here the question can be raised whether mankind benefits from knowing the secrets of Nature, ...The example of the discoveries of Nobel is characteristic, as powerful explosives have enabled man to do wonderful work. They are also a terrible means of destruction in the hands of great criminals who are leading the peoples towards war. I am one of those who believe with Nobel that mankind will derive more good than harm from the new discoveries." As for criminals, small or large, there are many in our midst. Once again, we hope for the best.

Let us end with a last information. Since many years Mr. Rolando Pelizza is claimig to be in possession of a *machine* able to perform incredible tasks [a machine that received in due time favorable but *indirect* evaluations by scholars as the Italian professor of physics Caglioti, former president of CNEN]. Mr. Pelizza claims also that various foreign authorities keep him from using his "machine" since he pretends it to be employed for peaceful purposes ONLY.

But let us forget about the machine itself:

> The point is that Mr. Pelizza asserts that machine to be the invention of Ettore Majorana, whom he contacted in a secured place many years after his disappearance,

and for a long time...; and that it was Majorana to strictly impose the condition of an exclusively peaceful use of it. Mr. Pelizza produced also letters, supposedly written when Majorana was 60 or 90; and photographs; together with some experts' reports obtained by him. About Majorana we don't care about "experts' reports"; nor we are experts on photos... But we may say that the handwriting, the calligraphy, appears to be his own (even too much). However, the content of those letters — two of them addressed to us — are *not* compatible at all with Majorana's style. Therefore, the whole story has not convinced us (even if it could furnish a justification for Majorana disappearance…); and therefore we are mentioning it quite briefly and only *en passant.*

Handwriting Analysis

Perhaps to help understand Ettore's emotional state at the time of his disappearance, an examination of his handwriting could be revealing. An educated friend, historian, and incognito essayist, who also secretly delights in graphology, agreed in 1972 to study a sample of Majorana's writing. Here is Dr. Gianni Sansoni's response: "Dear Erasmo, never before this moment has graphology revealed itself to be so useless, even disrespectful. I think that the recommendations of Cesare Pavese can easily be adapted to Majorana: do not engage in gossip. But when has there ever been respect for the solitary?… I say this because last night's conversation has reinforced my previously-formed conviction that first and foremost Ettore Majorana was (and I don't exclude that he still is) of very high moral character and that he must have suffered greatly... In any case, my first impression is that he is *restless, never satisfied with himself or with others.* What worsens the picture is the second finding, which is that his soul *does not seem anchored in a divine entity or belief.* Quite the contrary, I note in the use of certain letters a *reluctance to see over and above things.* I reconfirm that the subject is fundamentally a positivist, and gifted with a *stringent, consequential logic.* The third "unpleasant" surprise: *introversion,* which is fed by a vast background of *skepticism* and *pessimism.* In short, we have up to this point a picture of a person who is somewhat "gattopardian", influenced by a well-defined class and secular position. Do you understand? Natural logic and *strong will* combined with skepticism will probably force him into genuine doubts and torments which he will not wish to overcome at all if he cannot do so using logic. A tragic closed circle that only a purpose, a goal, an authentic teleology could dissolve. Allow me to make a few comments on the overall appearance of the letters. Very good style, elegant, cultured, which fits his physical appearance. They also reveal a love for precision, a strong, deep and ineradicable desire for clarity and logic. And is the quote about the "Ibsenian girl" just a cultural habit, or deep down does it reveal that he fears the discovery of his desire for renewal? We talked about the sea that "refused" him and of the expression "I will hold dear memories, at least until eleven o'clock"; a see-saw of emotions (the tango songs of that age are full of the boy and girl who come, go, return, run away,...) or rather a sense of panic over a deadline, of a duty from which he cannot escape? As you see, we cannot pore over someone without starting a discourse that will be unavoidably too long... But I can say this: Majorana must have been a *gentle and good person, in need of affection more than ever*, and I think the best eulogy one could give him is to approach his situation with respect and understanding."

Epilogue

We have attempted to understand some of the possible "reasons" for Majorana's life.

We have also shown the reader, the final judge, the known evidence supporting a suicide, and that (more substantial in our opinion) in favor of an escape from his own Pirandellian *puppets,* to a place where he could start a new life — one more balanced, more his own, more real. There is just one piece of information that we must, in fairness, add: While director Salvo Ponz de Leon, during a brief visit to Buenos Aires, believed he had found evidence of the Argentinian track, in director Bruno Russo's 10 days of investigation in Argentina (together with mutual friends Roberto Ferrari and Luiz

Bassani) no confirmation of this was found, apart from what has been described in this chapter. That is, Bruno Russo found no trace of Ettore Majorana, Mrs. Talbert or Tullio Magliotti at the registry office or in the telephone directories. Leonardo Sciascia, however, with his probing spirit, continued to believe that some of the reports concerning Argentina were reliable (letter *T/A8*).

We will now yield the floor to someone who did already express her opinion: the Italian critic Aurora F. Bernardini, who works in San Paolo, Brazil. She said, "This credible and well-grounded hypothesis on the survival of Majorana is not only more generous but more revolutionary (or at least more progressive) than a convenient suicide... *Discarding* completely the common belief that the geniuses of physics are precocious and short-lived, or that a physicist can possess great talent in his field and be an imbecile in the rest, based on what we know of Majorana we cannot help but believe that his genius led to the discovery of *his* truth. Or of the truth *tour court*, that Tolstoi's Ivan Ilych discovers just before dying. What are the moments in life when we are truly alive? Everyone finds their own answer, almost always belatedly. Probably Majorana found his early. His legacy in this regard would be very useful for humanity today. Maybe even more useful — *honni soit...* — than his legacy as a physicist."

Napoli, 25 marzo 1938-XVI

Caro Carrelli,

Ho preso una decisione che era ormai inevitabile. Non vi è in essa un solo granello di egoismo, ma mi rendo conto delle noie che la mia improvvisa scomparsa potrà procurare a te e agli studenti. Anche per questo ti prego di perdonarmi; ma sopra tutto per aver deluso tutta la fiducia, la sincera amicizia e la simpatia che mi hai dimostrato in questi mesi.

Ti prego anche di ricordarmi a coloro che ho imparato a conoscere e ad apprezzare nel tuo Istituto, particolarmente a Sciuti; dei quali tutti conserverò un caro ricordo almeno fino alle undici di questa sera, e possibilmente anche dopo.

E. Majorana

Letter written to the director (Prof. Carrelli) of the Physics Dept. of the Naples University, in which Ettore Majorana announced his decision to disappear. Naples, March 25, 1938. (*Reproduction forbidden.*) Photo courtesy of M. Majorana and E. Recami.

Ettore Majorana (on the right), as a child, with his two brothers Luciano and Salvatore. Photo courtesy of E. Recami. *(Reproduction forbidden.)*

Ettore Majorana at six years old. Photo courtesy of E. Recami. *(Reproduction forbidden.)*

Ettore Majorana's grandfather (Salvatore Majorana Calatabiano), the "progenitor" of the Majorana family, who became twice a minister in De Pretis' government of Italy. Photo courtesy of E. Recami. *(Reproduction forbidden.)*

Ettore Majorana's uncle Prof. Angelo, who became a minister twice in Giolitti's government of Italy, besides also being the rector of the Catania University. An extremely brilliant, and precocious scholar. Photo courtesy of E. Recami. *(Reproduction forbidden.)*

Ettore Majorana's uncle, Prof. Dante (who was also Rector of the Catania University), with his wife Sara and his son Salvatore. Photo courtesy of E. Recami. *(Reproduction forbidden.)*

Ettore Majorana's mother, Dorina, with her three children, including Luciano (Catania, 1908). Photo courtesy of E. Recami. *(Reproduction forbidden.)*

Ettore Majorana's uncle Prof. Quirino, a renown and brilliant experimental physicist, who acted also as the President of the Italian Physical Society from 1926 to 1946. Photo courtesy of E. Recami.

Ettore Majorana's mother, Salvadora (Dorina) Corso, when young. *(Reproduction forbidden.)*

Ettore Majorana's father, Dr. Eng. Fabio Massimo, when young. *(Reproduction forbidden.)*

Ettore Majorana's grandmother and grandfather (the progenitor of the Majorana family, twice a minister in De Pretis' government of Italy), with all their sons and daughters: Angelo, Giuseppe, Quirino, Dante, Fabio M. (the father of Ettore Majorana), Elvira and Emilia. Three of the sons became Rectors, in different times, of course of the same University, and one of them also a minister twice, in, Giolitti's government of Italy. *(Reproduction forbidden.)*

Ettore Majorana (*centre*) in a boat (Porto Santo Stefano, Italy; 3 January 1926) with his brother Luciano (second from the right). Photo courtesy of E. Recami. *(Reproduction forbidden.)*

Ettore Majorana (the first on the left), during an excursion on the Etna mountain (December, 1926) with friends and his brother Luciano (first on the right). Photo courtesy of E. Recami. *(Reproduction forbidden.)*

Ettore Majorana at Abbazia (at that time in Italy, now in Slovenia), a famous sea resort, with his sisters Rosina and Maria (August 1932). Photo courtesy of E. Recami. *(Reproduction forbidden.)*

Ettore Majorana at Abbazia (at that time in Italy, now in Slovenia) with his father and his sisters Maria and Rosina. July 1931. Photo courtesy of E. Recami. *(Reproduction forbidden.)*

Ettore Majorana at 17 years old (3 November 1923) , as a first-year university student.

A nice portrait of Ettore Majorana at 23 years old. Photo courtesy of E. Recami. *(Reproduction forbidden.)*

Ettore Majorana: a portrait, in profile. Photo courtesy of E. Recami. *(Reproduction forbidden.)*

Ettore Majorana: a portrait. Photo courtesy of E. Recami. *(Reproduction forbidden.)*

Ettore Majorana as a high-school student, signed by him. Photo courtesy of E. Recami. (*Reproduction forbidden.*)

B. Pontecorvo, E. Segre' and E. Amaldi, together again in Rome after 40 years (in 1978); signed by them on the spot. Photo courtesy of E. Recami.

The Fermi's group library (Department of Physics, Via Panisperna; Rome), with a table, also used by Ettore Majorana. Photograph courtesy of E. Recami.

The main lecture hall of the Naples Department of Physics (at the time in Via Tari 3), where Majorana exerted his full-professorship "for exceptional merits".

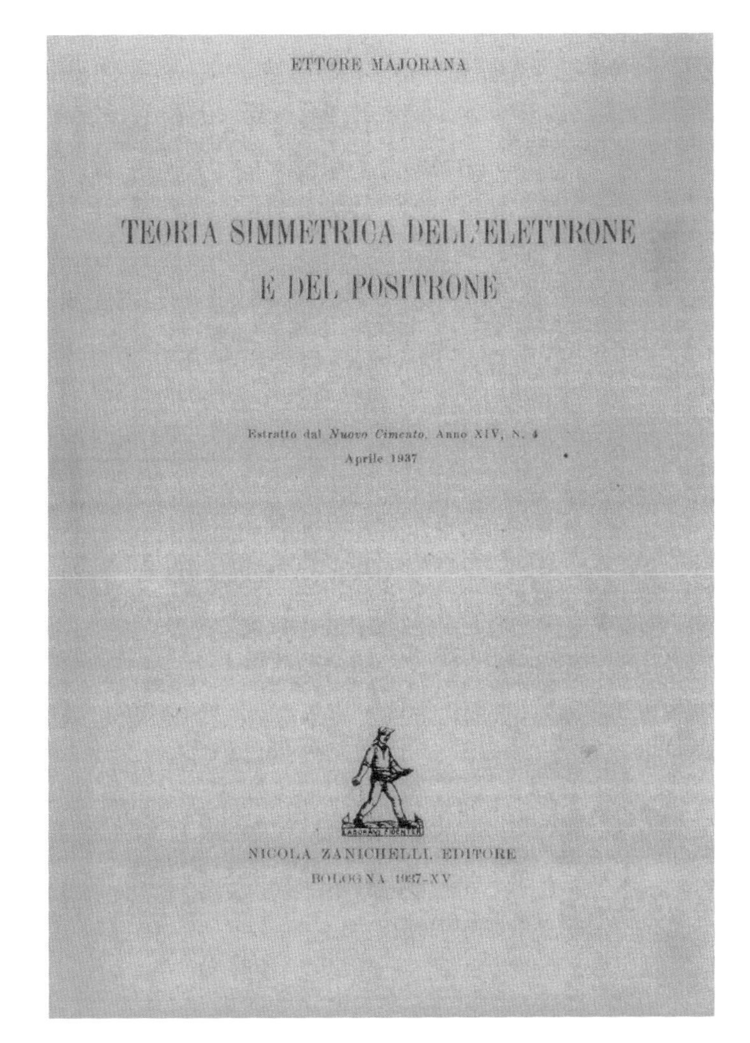

ETTORE MAJORANA

TEORIA SIMMETRICA DELL'ELETTRONE E DEL POSITRONE

Estratto dal *Nuovo Cimento*, Anno XIV, N. 4

Aprile 1937

NICOLA ZANICHELLI, EDITORE

BOLOGNA 1937-XV

First page of the *Reprint* of the most famous publication of E. Majorana, on Majorana's neutrino and Majorana fermions.

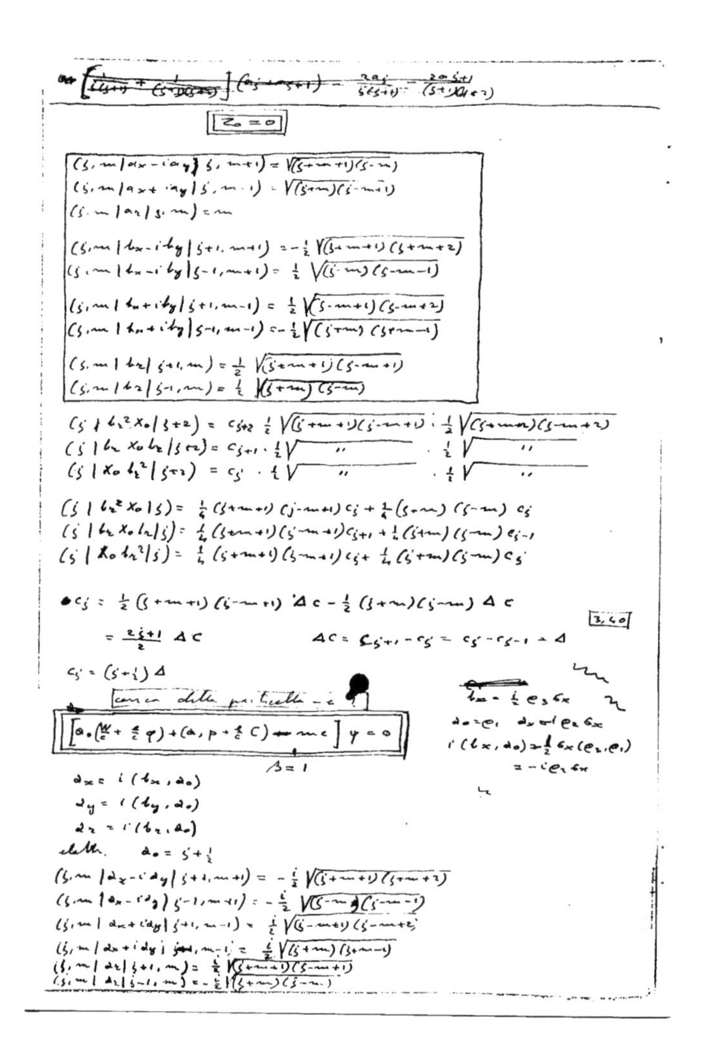

Some of Ettore Majorana's notes in preparation of his infinite-component equation paper [manuscript deposited by E. Recami & M. Majorana] at the Ettore Majorana Center for Scientific Culture of Erice (TP, Italy). Photo courtesy of E. Recami.

F. Rasetti, E. Fermi and E. Segre', in their academic gowns in Rome. It later became known as "the photo of the three priests" (explicitly mentioned in a F. Rasetti's letter reproduced in the text).

Bruno Pontecorvo

A photo of Bruno Pontecorvo, a member of the Fermi's Group at Rome (together with E. Majorana).

Ettore Majorana (*center*) in Viareggio's pinewood, Italy, August 1926, together with, from the left, his mother, his sisters Maria and Rosina, his friend and fellow student Gastone Pique', and his maternal grandmother. Photo courtesy of B. Pique' and E. Recami. *(Reproduction forbidden.)*

Ettore Majorana (the second from the right) together with his brother Salvatore, his mother (*center*), and his sisters Maria and Rosina in Karlsbad, at that time Czechoslovakia, during autumn of 1931. Photo courtesy of E. Recami. *(Reproduction forbidden.)*

A group of gentlemen in the country: The arrow indicates Ettore Majorana's father. Photo courtesy of E. Recami. *(Reproduction forbidden.)*

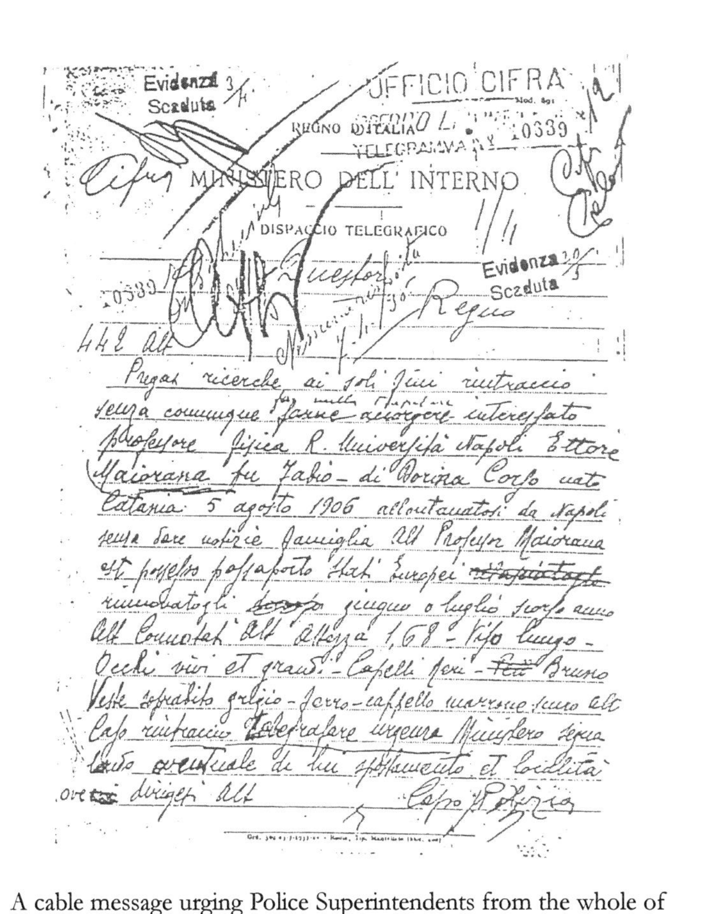

Evidenza 3/4
Scaduta

UFFICIO CIFRA

REGNO D'ITALIA

TELEGRAMMA

MINISTERO DELL' INTERNO

DISPACCIO TELEGRAFICO

Evidenza
Scaduta

Questori Regno

442 Alt

Pregasi ricerche ai soli fini rintraccio senza comunque farne accorgere interessato professore fisica R. Università Napoli Ettore Majorana fu Fabio – di Dorina Corso nato Catania 5 agosto 1906 allontanatosi da Napoli senza dare notizie famiglia Alt Professor Majorana est possesso passaporto Stati Europei rinnovatogli giugno o luglio scorso anno Alt Connotati Alt altezza 1,68 – viso lungo – Occhi vivi et grandi – Capelli neri – Bruno Veste soprabito grigio – ferro – cappello marrone scuro Alt Caso rintraccio telegrafare urgenza Ministero segnalando eventuale di lui spostamento et località ove dirigesi Alt

Capo Polizia

A cable message urging Police Superintendents from the whole of Italy to increase their search of Ettore Majorana. It was written by the Chief of the Italian Police, Senator Bocchini, on March 31st, 1938, and sent on the same day to the "Cipher Office" with no.10639. All documents of this kind come from the Italian Ministry of the Interior (Division of General and Classified Affairs), and are now stored at the State Central Archive, Rome.

Sculpture in bronze (Ettore Majorana's bust), by the artist Carlos (he appears behind the bust) D'Agostino, located at Passo Piasciaro, on the Etna mountain (where Ettore Majorana used to spend his summers). Photo by La Spina. Courtesy of E. Recami.

BOLOGNA, 27/4/1984

Signatures collected by E. Recami during the meeting "Fifty Years of Weak Interactions" (Bologna, Italy, April 1984). One can recognize, e.g., the signatures of Lucio Mezzetti, Carlo Rubbia, Nicola Cabibbo, Bruno Pontecorvo, Oreste Piccioni, Antonino Zichichi, Antonio Rostagni, Gilberto Bernardini, Piero Caldirola, Carlo Castagnoli, Giuseppe Occhialini, Claudio Villi, Raoul Gatto, Giancarlo Wick, Franco Bassani, Ettore Fiorini, Francesco Fidecaro, Emilio Picasso, Luciano Maiani, Milla Baldo Ceolin, Alberto Gigli Berzolari, Giorgio Salvini, etc. Photo courtesy of E. Recami. *(Free to use with credit.)*

Maria Majorana, Ettore's sister (the second from the right), together with the author, on April 20, 1986, at a meeting about Majorana held at Mirano (Venice), organized by Q. Bortolato and with the presence of Abdus Salam (Nobel Prize Laureate), E. Amaldi, and the president of the Italian Phys. Soc., amongst others.

- 101|1 -
(quaderno 2)

$$4\pi\rho = \operatorname{div} E \qquad \operatorname{div} H = 0$$

$$4\pi j + \frac{1}{c}\frac{\partial E}{\partial t} = \operatorname{rot} H \qquad -\frac{1}{c}\frac{\partial H}{\partial t} = \operatorname{rot} E$$

$$\psi_1 = E_x - iH_x$$
$$\psi_2 = E_y - iH_y$$
$$\psi_3 = E_z - iH_z$$

$$\operatorname{div}\psi = \operatorname{div} E - i \operatorname{div} H = 4\pi\rho$$

$$\operatorname{rot}\psi = \operatorname{rot} E - i \operatorname{rot} H = -\frac{1}{c}\frac{\partial H}{\partial t} - \frac{i}{c}\frac{\partial E}{\partial t} - 4\pi i j = -\frac{i}{c}\left(\frac{\partial E}{\partial t} + i\frac{\partial H}{\partial t}\right) - 4\pi i j$$

$$4\pi j + \frac{1}{c}\frac{\partial \psi}{\partial t} = i \operatorname{rot}\psi$$

$$\frac{1}{c}\frac{\partial \psi}{\partial t} - i \operatorname{rot}\psi + 4\pi j = 0$$
$$\operatorname{div}\psi - 4\pi\rho = 0$$

$$\frac{1}{c}\frac{\partial \psi_1}{\partial t} - i\frac{\partial \psi_3}{\partial y} + i\frac{\partial \psi_2}{\partial z} + 4\pi j_x = 0$$
$$\frac{1}{c}\frac{\partial \psi_2}{\partial t} - i\frac{\partial \psi_1}{\partial z} + i\frac{\partial \psi_3}{\partial x} + 4\pi j_y = 0$$
$$\frac{1}{c}\frac{\partial \psi_3}{\partial t} - i\frac{\partial \psi_2}{\partial x} + i\frac{\partial \psi_1}{\partial y} + 4\pi j_z = 0$$
$$\frac{\partial \psi_1}{\partial x} + \frac{\partial \psi_2}{\partial y} + \frac{\partial \psi_3}{\partial z} - 4\pi\rho = 0$$

$$\frac{W}{c}\psi_1 + i p_y \psi_3 - i p_z \psi_2 = 0$$
$$\frac{W}{c}\psi_2 + i p_z \psi_1 - i p_x \psi_3 = 0$$
$$\frac{W}{c}\psi_3 + i p_x \psi_2 - i p_y \psi_1 = 0$$
$$p_x\psi_1 + p_y\psi_2 + p_z\psi_3 = 0$$

$$\left(\frac{W}{c} + \alpha_x p_x + \alpha_y p_y + \alpha_z p_z\right)\psi = 0$$

$$\alpha_x = \begin{vmatrix} 0 & 0 & 0 \\ 0 & 0 & -i \\ 0 & i & 0 \end{vmatrix} \quad \alpha_y = \begin{vmatrix} 0 & 0 & i \\ 0 & 0 & 0 \\ -i & 0 & 0 \end{vmatrix} \quad \alpha_z = \begin{vmatrix} 0 & -i & 0 \\ i & 0 & 0 \\ 0 & 0 & 0 \end{vmatrix}$$

$$\alpha_x\alpha_y - \alpha_y\alpha_x = i\alpha_z$$

A page extract of a scientific manuscript (published by us in 1974) in which Ettore Majorana in 1931 started deriving a Dirac-like equation for the photon. Majorana had left unpublished a great majority of his scientific manuscripts.

Part II

Letters, Documents, and Testimonies

9

Letters and Testimonies

T/1 — Testimony of Gastone Piqué (28th July 1984)

Rome, July 28, 1984

Dear Professor Recami

I am very grateful for this opportunity to recall my dear friend Ettore, the memories of whom date back to when I was a young man. I would also like to thank you for allowing me to make a contribution, albeit small, to the work you are conducting in his memory.

My recollections range from high school (Liceo Tasso in Rome) to the second last year of the School of Applied Engineering, which Segré convinced Ettore to leave in order to continue his studies under Enrico Fermi; Segré himself had done the same a few months earlier, leaving the Institute of Engineering to switch to Physics in Via Panisperna. The memory of Ettore and his affectionate friendship during our youth shall always remain dear to me.

I have no problem at all with sending you the photocopies of three letters and a postcard I received from Ettore. This is all that remains of our intense summer correspondence from 1925 to 1928; a correspondence that was only ever jovial and brotherly. It reveals the spirited and caustic moods of Ettore which many others who subsequently spoke of him denied. After Ettore dropped out of engineering studies, our friendship continued for another two or three years until I was hired by 'Società Romana di Elettricità' following my degree and Ettore was appointed professor at the University of Naples.

I believe you will find the short annotations to the attached correspondence useful.

1st letter: Ettore wrote to me after the death of my father and the affection he expressed in this letter is particularly dear to me.

2nd letter: is interesting for the comical description he gives of his stay in Montecatini, where the Majorana family in this period used to spend their summers before briefly stopping in Viareggio, where I spent my summer vacations.

Postcard: in this postcard Ettore announces his arrival in Viareggio, indicating where he hoped to meet me: at no. 42 Via Vespucci [actually, it was via Buonarotti], *where I lived; at the "Gatto Nero" café, located in the pine West grove which I habitually frequented; at the thermal bath "Felice", where I and my family had rented a cabin; "beneath those windows" by this alluding to the house of a young girl whom I was courting at the time. Again in this postcard you get a sense of Ettore's 'humor'.*

3rd letter: was written in October 1927 from Passopisciaro; the Majorana family went there every year in that period for the grape harvest. It is interesting to note his skeptical and sardonic language in this letter, and his bitter yet ironic self-proclamation as an "immature genius".

What I have written is in the memory I will always have of Ettore, and the affection that we shared…

You have my permission to use the photocopies as you see fit, in the hope that future generations will preserve the memory of the amazing Ettore; for he was amazing even to his intimate circle of loved ones and boyhood friends.

It would give me great pleasure to see you and talk together about Ettore the next time you are in Rome. My kindest regards

Gastone Piqué

Dr. Engineer Gastone Piqué
Rome.

MP/1 — First Letter to G. Piqué (4th September 1925)

Rome, September 4, 1925

Dear Gastone,

I feel a deep pain in my soul for this irreparable disaster that has struck one of my dearest friends.

You, one who has never shirked sacrifice and work, have the capacity to calmly and bravely cope with this premature, albeit long-feared, passing of your unforgettable father, and carry on under the burden of your new and serious duties. Anyone who, like me, has ever come close to your kind family would have no difficulty in sensing the complete and touching harmony and the perfect unity of purpose you had; now that tragic fate has delivered desolation, use the memory of the undisturbed happiness that was, to relieve your pain and give you strength.

In a few days I will go to Sicily; I hope that I will see you on your next visit to Rome before you leave.

On behalf of my family, and especially myself and Luciano, I assure you and your family that we all actively share in your pain.

Believe me

Yours very affectionately, Ettore Majorana.

MP/2 — Second Letter to G. Piqué (2nd August 1927)

HOTEL TAMERICI
MONTECATINI BAGNI

August 2 [1927]

Dear Gastone,

I stoically drank three drops of bitter water, and then another ten drops, and then another six glasses. I am waiting; may God help me…

The languor which pervades my soul induces in me the most tender of feelings. What gentleness in the respite of one who has sipped a liter and a quarter of purgative water! … The delicate charm of your beautiful land and the people who inhabit it, along with a subtle sense of nostalgia, complete the enchantment.

I hope to see you again in three days, perhaps tomorrow if I come to reserve the rooms. I am looking forward to that moment; because there is no greater joy for me than to see you.

Ettore

P.S. Duty calls… Goodbye!

MP/3 — Postcard to G. Piqué (4th August 1927)

To Mr.
Gastone Piqué
Via Buonarroti 42
Viareggio

August 4 [1927]

Dear Gastone,

I will come perhaps tomorrow, in the afternoon. I'll look for you actively and methodically:

1st) at no. 42
2nd) at the Gatto Nero
3rd) at the Felice lido
4th) beneath those windows.
Greetings

Ettore.

MP/4 — Third Letter to G. Piqué (17th October 1927)

Passopisciaro, 17.X.1927

Dear Gastone,

I thank you sincerely for your usual efforts; I have not written up to now because I do not like to rush, especially in certain things. You should know that I have engaged in the most scientific of pastimes: I do nothing and time passes just the same. Actually I am working on an incredible number of things, but being vile matters of thought, and not empirical facts, we must not give too much credit to them.

If no misfortune befalls me, I'll be there in a few days. Nor must you believe that it is impossible that it could happen to me in the prime of my life. On the contrary, I reckon it to be quite likely. In fact, from birth I've been a stubbornly immature genius; "time and straw" did not help and never will, and nature would not be so evil as to have me die prematurely of arteriosclerosis.

But though vast and unfathomable is the sea of my disregard for the entire sublunar world, it is not without joy that I prepare myself to cross the threshold of the famous hall of Via Montecatini, nor is it without trepidation that I will drink the bitter chalice, till the last drop.

Affectionately,

Ettore.

MG/R1 — First Letter[1] to Giovanni Gentile jr. (22nd December 1929)

Rome, 22.XII.1929 – VIII

Dear Gentile,

Thank you very much for your first and interesting news from Berlin, which I received only a few days ago. As

[1] This first letter, as well as the fourth, fifth, seventh, eighth, ninth and tenth letters to G. Gentile jr., were discussed — with some degree of inaccuracy — by B. Gentile in the *Giornale Critico della Filosofia Italiana* (1988); while the sixth was released to the public by E. Gentile and first appeared in an article by Piero Bianucci (*La Stampa*, 28.5.1989). The citation below of Landau refers to *Physikalische Zeitschrift*, 30 (1st October 1929), p. 654.

soon as I have confirmation of your new address, I shall send you some of the new works by Fermi; I must also respect the wishes of the illustrious [Johann] Kudar and send you the *only* copy of my thesis, even though it contains very little of interest.

I read [George] Gamow's article that you brought to my attention, and I think that it certainly provides insight to the first cries of the nascent theory of nuclei. I feel, however, that it will never mature unless it is grafted onto the trunk of quantum electrodynamics, which in turn emits still more pitiful cries (read z.B. [= e.g.] an article which I believe is by [Lev D.] Landau, in *Physikalische,* on or around November 1).

In other words, the problem of the grouping of protons and electrons in nuclei[2] seems unlikely to be resolved, even on an approximate basis, if the problem of the constitution of these same protons and electrons is not first resolved. And this is for a very simple reason: the size of complex nuclei are, according to the "Gamoviche" theories, are of the same order of magnitude as the size of the electrons (calculated classically, of course. Quantum theory has not, and *cannot alone*, shed any light on the matter as there cannot be, for dimensional reasons, any relationship between

---

[2] It was believed at the time (neutrons having yet to be discovered) that atomic nuclei were made up of protons and electrons. Majorana thus encountered theoretical difficulties — confirmed in his unpublished scientific manuscripts — in describing nuclei on the basis of this hypothesis.

e, *h* and *m* or, what is the same, between *e*, *h* and the "radius" of the electron). Given that these statements are enormously vague, it is presumable that there is some underlying truth. Experience that endures uninterruptedly since the dawn of quantum theory warns us to never sneer at suggestions that may derive from certain parallels with classical theories. For this reason, I tend towards the belief that protons and electrons "penetrate each other" in the nuclei in a manner that is not associated with wave mechanics, and cannot therefore be subject to statistical interpretation. None of us would hesitate to swear "a priori" on the contemporaneity of the Compton Effect, even though the analytical explanation of it is highly intricate in my opinion. Even here, what drives us is our *faith* in classical analogies.

Here they are working on molecules and hyperfine structures, both in theory and in practice. Fermi has also solved a problem we talked about last time, that is, the anomalies in the intensity of the absorption doublets of the alkali. He showed that the anomalies are only apparent, and simply derive from the inadequacy of the calculation of first approximation (which results in a 2:1 ratio of intensity between the lines of a doublet). They occur for higher lines, and never for the first, which is by far the most intense.

As for me, I am not doing anything sensible, that is studying group theory with the firm intention of learning it; similar to Dostoevsky's hero who one fine day began putting a few coins aside, with the idea of soon becoming as rich as Rothschild.

I hope my warmest Christmas wishes fly off this sheet of paper and race ahead of it to arrive on time. Best regards and best wishes also on behalf of Fermi, Rasetti and Segré, and the whole crowd of the friend and admiring physicists, including myself.

Ettore Majorana.

MG/R2 — Second Letter to G. Gentile jr.[3] *(15th May 1930)*

Rome, May 15, 1930

Dear Gentile,

I received your postcard from Florence, followed by your lovely letter from Leipzig. For both of which I thank you kindly. I sent the thesis to Bernardini.

Rome does not offer me, the way Leipzig offers you, any wondrous novelties, nor does it provoke under its vast, hazy sky so vast a crop of serious thought. Causes for deep reflection are rare here and even fewer those who attribute them any real weight. And yet this is the land blessed by God. The thousand German sages shine like beacons in the fog, rivaling but never besting our blazing sun in bestowing light and warmth on poor mortals. And if ever the artifice should here second our prodigal nature (if we have the

[3] In this wonderful letter, Ettore explains to Giovannino Gentile in his usual eloquence that the news that he received from Germany is incorrect, but he already has the correct answers (and will send them to him in his next letter).

"big guns" and, above all, a little bit of seriousness) the laurels of victory shall flourish in this immortal land. I do not say this to diminish our respect for a tenacious race, but because it be not without intention our regarding the work of others as well as the fruitless attempts of others, because what fails elsewhere is destined to triumph under a more amicable sky. Moving to matters that are closer to home, I can tell you that the pace of scientific activity in our institution is weakening, both due to the usual effect of the summer heat and the next departure of the "Pope", which, as you know, will take place on June 7.

I will read the *Berliner Tageblatt* of May 11, and one of these days I will see Pirandello's *As You Desire Me*, (Come Tu Mi Vuoi), which you described to me with such enthusiasm. As for the curve of potential for two helium atoms, you can extend it to wherever you need, since it is enough to calculate the first approximation terms in one or two distant points and then interpolate exponentially, as Slater does, even down to small distances; for the second approximation terms due to polarization, you already have the asymptotic expression. I will certainly inform you of any important news and I'm looking forward to hearing even bigger news from Germany. Best wishes on behalf of the Pope, the Sacred College, the Seminarians and the Friars Minor, as well as myself, and I wish you the utmost success in this second part of your stay abroad.

Yours Ettore Majorana.

MG/R3 — Third Letter to G. Gentile jr. (19th May 1930)

Rome, May 19, 1930

Dear Gentile,

I hasten to respond to various puzzles regarding the forces of polarization. In our formula [...].

I saw *As you desire me* (Come Tu Mi Vuoi), which I very much enjoyed; some speak of the unbearable Berlin taint in the main actress. I was unable to find the *Berliner Tageblatt* of May 11, sold out everywhere.

Greetings and best wishes from everyone,

yours,
Ettore Majorana.

MF/R1 — Postcard from Rome to His Parents (24th December 1915)

Eng.
Fabio Maiorana[4]
Via Etnea 251
Catania

[*Rome, 24.12.1915*]

Dear parents

Greetings to all and Merry Christmas.
You will receive[5] a kiss from
your loving son

Ettore.

---

[4] *Sic.*

[5] *Sic.*

MF/R2 — Letter from Rome to His Father (n.d.)

CONVITTO "MASSIMO"
ALLE TERME
ROME

Dear father, I received a letter from mother. When will you come? We are all well.[6]

I have not received your letters. How are you and mother? Has grandmother returned? Will Maria also come?[7] It rains here quite often. A million kisses. Your loving son

Ettore.

MF/R3 — Letter from Rome to His Mother (n.d.)

Dear Mother

We are all well. When will father come? Will you come for Carnival? We often cannot go for walks due to bad weather. Please also bring grandmother to Rome. Did you get my letter? Valentino [Dominedò] has recovered and has been well for many days.

It seems to me that in the letter I sent you I wrote all that you asked me. If you want to know anything else, write me immediately. A million kisses. Your loving son

Ettore.

---

[6] Also referring to his brothers and cousins: see Chapter 3

[7] His little sister.

MF/R4 — Postcard to His Mother (19th July 1916)

Mrs. Dorina Majorana
Via Etnea 251
Catania

Strada, 19-VII-1916

Dear Mother,

Dear Mother, we are all well and enjoying ourselves very much.
Please send me [the book] *"La guerra sul mare" [The War at Sea]*.
Greetings and kisses from
Your loving son

Ettore.

Ms/1 — "Information on My Didactic Career" (May 1932)

I was born in Catania, August 5th, 1906 I earned a high school (Liceo) diploma in classical studies in 1923; I then regularly attended engineering studies in Rome until the beginning of the final year.

In 1928, wishing to study pure science, I requested and obtained admittance to the Faculty of Physics, and in 1929 I graduated with a degree in Theoretical Physics under the direction of S.E. Enrico Fermi, writing my thesis on *Quantum theory of radioactive nuclei* and obtaining full marks and honors.

In the years that followed, I freely attended[8] the Institute of Physics in Rome following the scientific movement and carrying out various issues of theoretical research. I have continuously benefited from the wise and dynamic guidance of S.E. Professor Enrico Fermi.

Ettore Majorana
Rome, May 1932.

MX/R1 — Reply to an Invitation to Leningrad (1932?)

Rome [*1932 or 1935*]

Cher Monsieur

Je vous remercie vivement pour vôtre invitation de prendre part au pro-chain congrès qui aura lieu à Leningrade. Je suis heureux d'accepter et d'a-voir l'occasion de connaître à la fois votre grand et beau pays. J'ai parlé aussi de vôtre invitation à M. Fermi et à M. Rossi. Fermi est engagé pour un cours de conférences en Amérique et il ne pourra pas venir. Monsieur Rossi par contre m'a assuré qu'il acceptera très volontiers de se rendre en Russie.

Avec mes vifs remerciements recevez, cher Monsieur, mes salutations empressées.

Vôtre

Ettore Majorana

---

[8] I.e., "voluntarily": without getting any salary. See the accompanying notes for this *Information* in Chapter 5. The title of "His Excellency" (S.E.) refers to Fermi owing to his membership in the Accademia d'Italia.

Ettore Majorana
Physics Institute of the University of Rome
Rome

Bruno Rossi
Physics Institute of the Royal University
Padua.

D/CR2 — Letter from E. Fermi to CNR [National Research Council] (2nd January 1933)

ROYAL ACADEMY OF ITALY

Rome, January 2 1933/XI

Dear Bordoni,[9]

to confirm what I have previously told you, please be informed that, for several years now, Dr. Ettore Majorana has undertaken questions concerning the theory of the atom, and particularly the application of group theory in this field of research.

The brilliant results that he has already achieved in this field and his truly unique attitude to his targeted research give ample reason to believe that he would benefit greatly from a period abroad, during which he may continue his research.

With kindest regards and best wishes

Enrico Fermi

---

[9] At the time, the president of CNR [National Research Council] was Guglielmo Marconi and Giovanni Magrini was the secretary general. Its "National Committee for Astronomy, Physics and Applied Mathematics" was instead headed by A. Garbasso (president), U. Bordoni (vice-president), and E. Bompiani (secretary).

MB/R1 — First Letter to CNR (9th January 1933)

Rome, 09.01.1933-XI

Dear Professor Bordoni,

In relation to my grant application, via S.E. Prof. Enrico Fermi to the hon. National Research Council, I have the pleasure of communicating my planned work schedule as follows.

It is my intention to depart for Leipzig around the 15th of the current month and there occupy myself until the end of June, under the guidance of prof. W. Heisenberg, with theoretical research mainly involving the structure of nuclei and the relativistic formulation of the new quantum theory.

During the holiday period between the winter semester and the summer semester, I shall return to Italy for about fifteen days. I will spend the rest of the time participating in conferences and scientific meetings that are traditionally conducted in that period both in Germany and in Denmark. With best regards and thanks, and with my utmost respect

Ettore Majorana
R. Institute of Physics,
Via Panisperna, Rome.

MF/L1 — First Letter[10] *from Leipzig (20th January 1933)*

Leipzig, January 20, 1933
Physikalisches Institut
Linnéstrasse 5

Dear Mother,

I arrived late last night after an excellent trip. I had to stop for a few hours in Kufstein due to the chaotic manner in which Austrian currency control operates. At Brenner they did not register the checks made out in my name, and I was reproached in Kufstein for not having reported them. In the end, the episode was clarified before the police chief of Kufstein and they did not give me any more difficulties. I learned this is a rather frequent event.

I quite like it here. The city is beautiful and comfortable. Its inhabitants are very polite. I am staying temporarily at the Park Hotel, but I hope to find more a permanent solution quickly. You can address any mail, even those in the future, to the Institute of Physics, where I'll be spending most of my time.

I traveled from Florence to Bologna with Francesco [Dominedò]. At the Bologna station I saw Uncle Quirino,[11] Aunt Vincenzina and Carmela.

---

[10] The word "letter" (without the recipient) refers to "letter to the family".

[11] See Chapter 3.

Aunty told me that she will travel to Rome soon.

Give my love to everyone

very affectionate Ettore.

MB/L1 — Second Letter to CNR (21st January 1933)

Leipzig, 21.01.1933 – XI
Institut für Theoretische
Physik, Linnéstrasse 5

Dear. Professor,[12]

I have been speaking with the director of this institute, Herr Professor Heisenberg, regarding the activities to be performed. I look forward now to the elaboration of a theory for the description of particles with arbitrary intrinsic momentum, which I started in Italy, and regarding which I have given a summary report in the Nuovo Cimento (currently being printed). Regarding other minor activities that expect to complete by the end of next month, I will provide details in due time.

With kind regards and profound respect,

devotedly Ettore Majorana

---

[12] All of the "letters to CNR" are addressed to Prof. U. Bordoni. They were only written for bureaucratic purposes and were therefore quite formal; the information that these provide in terms of the scientific activities of Ettore in Germany are, however, of the greatest interest. See, for example, this letter and that of 3rd March 1933.

MF/L2 — Second Letter from Leipzig (22nd January 1933)

Leipzig, 22.01.1933
Institut für Theoretische
Physik, Linnéstrasse 5
(or Physikalisches Institut;
single entry for institutions of
Theoretical and Experimental Physics)

Dear Mother,

I hope to receive good news from you soon. I am doing very well. I'm still at the hotel, but I hope that tomorrow I will be able to secure a good room in a private lodging. There are so many to choose from, that all I have to really do is decide sensibly. Everyone at the Institute of Physics has welcomed me quite cordially. I had a lengthy conversation with Heisenberg, who is extraordinarily polite and friendly. I am on good terms with everyone, especially with Inglis, the American I met in Rome and who now frequently keeps me company and acts as my guide. My German is improving visibly.

In a few days I will receive a visit from Bernardini, who resides in Berlin-Dahlem and has returned temporarily to Italy.[13]

---

[13] See, in addition to Chapter 4, the testimony that follows (Pisa, 1984), written by Gilberto Bernardini.

The weather is pleasant — somewhat colder than Rome, but without the wind. It often snows gently.

Life here is not expensive and the many cafés and night clubs are also cheap, with great music and carnival-like entertainment, and are very crowded on Saturday night.[14]

I often go to the cinema in order to accustom myself to German conversation. Shows are at a fixed hour and sometimes reservations are necessary. The feature movie is accompanied by tasteful fillers, more that you would find in Italy. Many abandon themselves without restraint to roaring laughter; a sign, among many, of the meager social subjugation characteristic of the Nordic people.

The central station is truly a marvel; it's the largest and most beautiful in Europe (including the new one in Milan, so I'm told).

The Institute of Physics is, along with several other places, in a charming position, a bit out of place between a cemetery and a lunatic asylum...

The tram service leaves a little to be desired; many lines but few trams, like Rome before the reform.

The internal political situation seems permanently catastrophic, but does not appear to be of much interest to the people. On the train I noticed the stiffness of an officer of the Reichswehr, who was alone in the compartment with me. He could not lie something on his luggage rack or make even the slightest movement without forcefully clicking his heels together. This rigidity was

[14] Refer to the testimony of Gleb Wataghin, at the end of Chapter 7.

evidently due to my presence, and in fact it appears that this exquisite but sustained courtesy to strangers is part of the Prussian soldier's spirit, because while he would feel disgraced if he had not rushed to light a cigarette for me, his demeanor prevented me from exchanging a single word with him apart from the compulsory greetings.

I wrote to father and I sent several postcards, even to Nitto. If any reprints of "Nuovo Cimento"[15] arrive, please send me only a few of them, ten at most, in an open registered envelope (stamped registered). You will find suitable envelopes on the first shelf of Luccio's bookcase. When will Turillo [*his brother Salvatore*] take the professorship examination? If he happens to go to the Ministry after January 27th, he can pick up my decree of professorship (from Borsi)

Hugs; my warmest greetings to all.

love Ettore.

T/2 — Testimony of Gilberto Bernardini (5th September 1984)

SCUOLA NORMALE SUPERIORE
Piazza dei Cavalieri

Pisa, 05.09.1984

Dear Recami

...Thanks to your copy of the letter from Ettore to his mother, I am grateful to have exhumed a buried memory of my scattered

[15] He is most likely referring to his "infinite components equation": see Chapter 5, and footnote 5 in Chapter 6.

meetings with him during the period that I spent in Berlin with Lise Meitner; but I do not think, with the fading of my memory, that I can write of it without certain undue additions of the imagination.

In Bologna[16] *I spoke with Bruno Pontecorvo and I think that he, better and more thoroughly than I, could write you the "letter of testimony regarding E. M." that you have amicably requested. In discussing it with Pontecorvo, some memories were revived. Among these is that, with Ettore, I avoided talking about physics because anything I could say would have been, for him, insignificant. As happened to me later with Pauli, I considered it easier for me and less trivial for him to talk about, for example, how great it was to be born after Michelangelo and Beethoven.*

I have not yet received what you published in '75. I do not know if you are preparing a new and different publication with Ettore's sister, Maria. If so, then allow me to suggest that you disregard the exceptional talents of Ettore as a physicist, to emphasize the things in him that could evoke his complex human spirituality, much more extensive and enlightened than what writers have conjured. This is apparent when you read "The life and work of Ettore Majorana" written by Edoardo Amaldi and published by the Accademia dei Lincei in 1966.

All I can add now are my thanks for the consideration that Maria Majorana and you have afforded me, and my kind regards.

G. Bernardini.

---

[16] See Chapter 4.

MF/L3 — Third Letter from Leipzig (7th February 1933)

Leipzig, 07.02.1933

Dear Mother,

I received your letter dated February 2 and the prints that Turillo sent. I am in good health now; and even the weather is much better, with spring temperatures. I have found decent accommodation not far from the institute.

Feenberg, another American physicist with whom I have become friends, has arrived from Rome. We understand each other quite well in German. Tomorrow the so-called "magnetic week" begins in Leipzig, which draws nearly all the physicists of Germany. I will see many old acquaintances again.

I intend to remain in Leipzig until the end of February. March and April are, in fact, the vacation periods here. I'll probably make use of the opportunity to go to Zurich to visit Pauli, one of the most famous scientists alive today. Inglis will be with me, as well as Bloch, a Swiss man I met here and who also possesses the virtue of speaking perfect Italian.

How are things in Catania? It is not clear to me whether Luccio [*Luciano*] is with you or whether he stayed in Rome.

I do not know if I will have the opportunity to see the Leipzig Fair. The opening date should be March 5, which is the Sunday before the first Monday in March according to tradition, but some say it may be moved to avoid coinciding with the day of the election.

Affectionate greetings to all. Don't forget Nitto and family.

I got a postcard from Nitto. Thanks.

love Ettore

MF/L4 — Fourth Letter from Leipzig (14th February 1933)

Leipzig, 14.02.33

Dear Mother

I received your letter from Catania. I am very pleased by the good news about your health and about your stay in Catania. I'm sorry that the condition of Mrs. Zappalà has worsened; I hope that she can overcome the illness.

A quite animated international congress of physics took place in Leipzig. I established personal relationships with several famous people, particularly with Ehrenfest, who obliged me to explain some of my work in minute detail and invited me to go to Holland.

On March 1st, I will go to Copenhagen rather than Zurich, as Switzerland, like Germany, closes its schools in March and April, while Denmark follows the Italian model. In Copenhagen I will see Bohr and others with whom I am already personally acquainted. It is, together with Leipzig, the most important center for theoretical physics in Europe.

My lodging is not very good, but there's really nothing better in Leipzig. And it is close enough to the institute. I only pay four marks. I have of course organized my washing through the landlady. I got a postcard from father who will soon be in Rome. The temperature has dropped again, but within tolerable limits. The weather is fine.

The environment at the institute of physics is very pleasant. I am on excellent terms with Heisenberg, Hund and everyone else. I'm writing some articles in German.[17] The first is ready, and I hope to eliminate some linguistic confusion during proofreading.

My warmest greetings to Rosina and Maria.

Ettore

MF/L5 — Fifth Letter from Leipzig (18th February 1933)

Leipzig, 18/02/1933

Dear Father,

I got your postcard from Milan, from which I deduce that I will probably find you in Rome.[18] A few reprints

---

[17] Of these articles he published only one, that of the nuclear "exchange forces" (Chapter 5); but it is of the largest interest to know that he had prepared others: see also the Letter MF/L5, as well as Letters MB/L1 and MB/L2.

[18] Remember that the Majorana family, who were originally from Catania, now resided in Rome. His father in that period (the last of his life: he died the following year) was working in Milan.

have arrived from Rome, but they are still with Director Debye, who received them due to the missing delivery name. I have had no news lately regarding the Turillo's professorship; is the Commissioner better now?

I wrote an article on the structure of nuclei that Heisenberg liked very much, even though it contained some corrections to his theory. I will also extend and publish, in German, my last article published in the Nuovo Cimento[19]; this work contains an important mathematical discovery, as I was able to confirm through a consultation with the Dutch Professor van der Waerden, who teaches here and is one of the leading authorities on group theory.

On March 1st I will go to Copenhagen to meet Bohr, the grandest inspirer of modern physics, though now a bit old and senile. He is still considered a deep thinker; he speaks in a mixture of mumbled languages, almost as if he feared that if he spoke more clearly, people would actually understand him. Every day he adds one word to his new work that is expected to be of decisive importance.

I will stay in København[20] until Easter, and then I will only come to Italy for a few weeks as I would like to return here again by the end of April.

---

[19] Referring to the 10 or so extracts that he had requested (Letter MF/L2). See Chapters 5 and 6.

[20] København: original Danish name for Copenhagen in English, as referred to at the end of Letter MF/L6).

Give my love to mother, grandmother, Rosina, Maria, Turillo and Luccio.

love Ettore.

T/GW/3 — Testimony of Gleb Wataghin (1975)[21]

In Leipzig, where Heisenberg worked, I met Jordan, Debye, Max Born who was just arriving there, and Ettore Majorana, a young man who seemed, as he in fact was, a true genius [...] The camaraderie, the friendship that exists between scientists... manifested itself, for example, in the way our scientific discussions took place, similar to sporting events. In Leipzig, we would gather for two-hour seminars from two to four in the afternoon. In the morning, theoreticians sleep. Afterward, we would play Ping-Pong in the library, on a table meant for students. I can say that the champion was Heisenberg. Then we would go to a pub, and perhaps play chess. We also played chess at the Institute of Physics. Because Heisenberg was one of the directors, nobody complained about our playing Ping-Pong or chess in the library, but this would have been unthinkable at that time in other institutes [...] The seminars were attended by people from all over the world. For example, I remember that once during a seminar held by Norzig

---

[21] This testimony is part of an interview (recorded on tape) given by Gleb Wataghin in 1975 at the Institute of Physics, which bears his name, of the State University of Campinas (State of Sao Paulo, Brasil). The interview is therefore in Portuguese. But, as it is a verbal testimony in a generally conversational style of language, only the translation into English is given here.

and one of his colleagues, they were forced into a very challenging discussion following questions asked by Heisenberg and Ettore Majorana [...].

I remember Majorana in particular who, according to the judgment of many and in particular Fermi, was an exceptional genius... He was ill, suffered from an ulcer, and consumed almost exclusively milk; he did not participate in sports or gymnastics; he often took long walks by himself. He was not very communicative. But every now and then we would meet him on Saturdays. He was very critical. He found that everyone he encountered was ill-prepared, or stupid... He spent a lot of time on statistical laws as they applied to nuclear matter... The exchange symmetry between protons and neutrons might be complete, including charge and spin, or regard only the charge or the spin. This had not been proposed or studied by others. And the exchange symmetry of the sole positions of protons and neutrons (without touching the spin) allowed the statistical comprehension of why nuclear material should have a constant density. This meant that Majorana's theory held a great advantage over that proposed by Heisenberg.

MF/L6 — Sixth Letter from Leipzig (22nd February 1933)

Leipzig, 22.02.1933

Dear Mother,

I haven't received any letters from Rome, but from your previous letters I imagine that you have already returned from Catania. I will probably leave for Copenhagen on the 1st of March, so you can send any correspondence to my new address: Copenhagen; 15 Blegdamsvej; University Inst. for Theoretical Physics (Denmark).

I'll just manage to avoid the winter fair that takes place from March 5 to 12. In this period, Leipzig becomes uninhabitable because of the huge influx of merchants, and all the prices, especially those of the hotels and lodgings, are doubled.

The weather is variable; it's quite cold and it snows frequently and extensively. I am very well and I have not had even the slightest cold for almost a month.

I greatly regret having to leave Leipzig, where I enjoyed such a warm reception, and will gladly come back in two months. But I can't do anything here during the holidays because the institute is deserted: in fact, it is completely closed.

In the last "colloquium", the weekly meeting attended by hundreds of physicists, mathematicians, chemists, etc., Heisenberg spoke of the theory of nuclei, and he had much praise for the work I've done here. [*Heisenberg and I*] have become good friends after numerous discussions and several games of chess. The opportunities for these come during the receptions he holds every Tuesday evening for teachers and students at the institute of theoretical physics.

I expect to leave Copenhagen a few days before Easter and go directly to Rome.

Give my love to everyone

Ettore.

MF/L7 — Seventh Letter from Leipzig (28th February 1933)

Leipzig, 28.02.33

Dear Father,

I received your letter dated the 22nd; I'm not sure if you're still in Rome. I'll probably stop in Leipzig for another two or three days, as I need to chat with Heisenberg. His company is irreplaceable and I want to take advantage of it for as long as he remains here.

The lengthy sessions of the three *Orestani* prove the candidate's seriousness. I imagine they have already successfully finished and the commissioners have agreed with each other to write a flattering report.

The weather here has thankfully returned to normal, which is about zero degrees during the day. Only a few days ago we hit a low of thirteen below zero.

I have already written my future address to mother. I will repeat it in case this letter is sent back to Milan: Copenhagen; 15 Blegdamsvej; University Inst. for Theoretical Physics (Denmark).

I have an old friend, Placzek, who was in Rome a year ago, in 1931–32, that is, last year. I do not know which route I'll take to Copenhagen yet; but it will certainly be the shortest. I do not think it will be very cold there by this time. I'll write you when I arrive.

Affectionate greetings,

Ettore.

MB/L2 — Third Letter to the CNR (3rd March 1933)

Leipzig, 03.03.1933 – XI

Dear Professor, as the holidays have begun here in Germany, I am leaving today for Copenhagen, where I will stay until April 15 with prof. N. Bohr. I have sent a paper on the theory of nuclei to the Zeitschrift für Physik. A manuscript of a new theory on elementary particles[22] is ready and I will send it to this magazine within a few days.

With profound respect and best regards,

Ettore Majorana

Copenhagen, Blegdamsvej 15
Institute for Theoretical Physics
(Denmark)

MF/C1 — First Letter from Copenhagen (5th March 1933)

Copenhagen, 05.03.1933

Dear Mother,

I arrived in Copenhagen yesterday after a very good journey. The weather is rainy and warm; it is generally

[22] Majorana is certainly referring to the work mentioned in the previous (21.1.33) letter to CNR: a new theory for which he had already given a *summary report* in the article *Relativistic theory of particles with arbitrary intrinsic angular momentum*. Recall that this "preliminary" article — which appeared in *Nuovo Cimento*, 9, p. 43, 1932 — is of great importance and thus understandable *how interesting* it would be to find a copy of the following manuscript, perhaps already written in German!, which Ettore refers to.

warmer here than in Leipzig thanks to the wind from the sea. I did not have time yesterday to go to the institute, where I am hoping good news from you has arrived. It is closed today, as it is Sunday. I am staying in a small hotel with meals included; I will probably try to find another one closer to the physics institute.

Due to the American crisis, foreign exchange trading is suspended. Yesterday I spent the whole afternoon trying to cash a check because of the widespread belief of an imminent general collapse of all gold based currencies. In the end I was able to receive an advance and the rest will be settled next week. From what I have gleaned from the Danish newspapers today, the dollar retains parity.

Copenhagen is an enormous city with good architecture. The population, equally intelligent and civil from the highest to the lowest strata, is cut from the same template. This is undoubtedly the secret of the much-prized Nordic civility... In the long run it must be boring.

Coming from Germany, one has the impression of leaving Europe and entering a colony of Eskimos. The sense of social distinction is entirely absent. Danish writing seems largely a mixture of German and English, and can be understood to a certain extent. I have not yet studied the grammar. Not everyone knows German; I even found a travel agency with an employee who did not understand German, but spoke Italian. English is probably the most diffuse language.

Life is pretty expensive and poorly organized. On the whole, I much prefer Leipzig. I'll be staying here 5 or 6 weeks, and then I will come directly to Rome.

Affectionate greetings,
Ettore

Copenhagen, Blegdamsvej 15
Inst. for Theoretical Physics.

MF/C2 — Second Letter from Copenhagen (7th March 1933)

Copenhagen, 07.03.1933

Dear Mother,

I have been given a letter from father and postcards from Francesco and Valentino [*Dominedò*]. I also have your letter from Catania and I have taken all your advice, except for the fur coat, which I feel I have no need for. Here the weather is better again and it is not at all cold. I'm in very good health.

I'm sorry about the insignificant failure of Turillo, who will have no trouble finding the appreciation he deserves when he becomes more familiar with the workings of universities.

Bohr is good-natured and likes the fact that I speak German worse than he, and is very concerned about finding me a guesthouse near the institute. I am on good terms with Møller and Weisskopf. Placzek has been invisible, as he has been busy writing his 'Handbuch' since time immemorial. He phoned me several times, still speaking in a good trasteverine [dialect]. This evening I will have dinner at his house, along with Weisskopf.[23]

I have not spoken of the political situation in Germany, as it had not seemed very serious. There has been some shooting at night even in Leipzig resulting from electoral agitation,[24] but no movement of a general nature. The elections are also forthcoming in Denmark. Huge processions of communists parade through the center of the city chanting and displaying placards, mainly against Mussolini and Hitler. They cause more laughter than consternation.

I will try to learn Danish according to the Levantine method. I found a little guidebook to help me; it teaches strangers the language and how to use it while they explore Copenhagen.

---

[23] See testimony (Cambridge, USA, 1984), by Victor Weisskopf. See also Chapters 5 and 6.

[24] Ettore sent this news when he was already in Denmark: obviously to avoid alarming his family (see also Chapter 6). The "Lilli" that he refers to at the end of this letter — and elsewhere — was Maria's dog.

Is father still in Rome? I think that it's time for him to return to Rome, for good. Greetings to all, including Ignazio and Lucia and the maid. How is Lilli?

love Ettore

Copenhagen, Blegdamsvej 15
Inst. for theor. Physics

T/4 – Testimony of Victor Weisskopf (16th May 1984)

Massachusetts Institute of Technology
Department of Physics
Cambridge, USA

May 16, 1984

Dear Dr. Recami:

I thank you very much for your letter and for sending me your most interesting article regarding the fate of E. Majorana. It was a tragic story, and it is also tragic that we do not yet have cleared up his case.

I am very glad that you have found a letter in which Majorana says that he had good relations with me. I must admit that my memory isn't good enough to recall any details of the meeting I had with him. I wish I could recall it. I have only a vague recollection that I did have a discussion with Majorana about the newest developments in quantum electrodynamics.[25]

Very best regards,

Sincerely yours
Victor F. Weisskopf

[25] See also Testimony *T/12* of GianCarlo Wick, Chapter 11.

P.S.: I would be grateful to you if you would transmit a copy of this letter to his family. VFW/don.

MG/C1 — Fourth Letter to G. Gentile jr. (12th March 1933)

Copenhagen, 12.03.1933

Dear Gentile,

I have been in Copenhagen for a week and I have fit in perfectly from the moment I arrived. There are many Germans at the institute: [George] Placzek, [Victor] Weisskopf, [Hans] Kopfermann; not so many Danes, such as [Niels] Bohr and [Jans Peter] Møller. They are all very nice people. Kopfermann is an experimental physicist who studies hyperfine structures. A remarkable person who looks like a boy and has fought in four years of war. Many speak of you in Leipzig.

Bohr has now finished his work written in collaboration with [Léon] Rosenfeld, which extends on the ideas of [Werner] Heisenberg regarding the uncertainty principle for electromagnetic field measurements. He has just left for a short trip to the mountains in the company of Heisenberg. Upon his return, he will resume his apostolate for propagating the Spirit of Copenhagen.

We had a short visit from [Wolfgang] Pauli, a very smart and likeable chap.

They had no news of the positive electron in Leipzig. Here they say[26] that it is a formidable quid pro quo. Positive electrons are ordinary electrons that return to the nucleus after completing a circle. This is also the opinion of [Ernst] Rutherford.

I received from Rome a copy of the great publication by Fermi and Segré, which will soon appear in the memoirs of the Academy. This will be followed by another great work by Fermi and Amaldi on statistical calculations. I have sadly received news of the death of Casare, the mechanic. There is little support for Hitler in Germany and Denmark and there are predictions, I consider unfounded, of his imminent fall. His first acts of government, particularly the total replacement of local administrations with nationalist elements, suggest that he knows what he is doing.

I will remain for about another month in Copenhagen. Then I will return to Rome for a few weeks, where I hope to see you again. How are you getting on in Pisa? I regret I have not seen Bernardini again; he left Leipzig when I had the flu.

Placzek will probably go to Rome between the end of April and beginning of May. And after a few weeks continue on to Russia by sea. Even [Hans A.] Bethe, who I saw again in Leipzig, said that he will be going to Rome next April. Møller is also planning a long stay in

---

[26] Erroneously.

Rome, subject to the decision of the Rockefeller Foundation. [Felix] Bloch will spend next winter with us. It seems that there aren't many theoretical physicists to choose from outside of Leipzig, Zurich, Copenhagen and Rome.

Affectionate greetings and see you soon,

Ettore Majorana

Copenhagen, Blegdamsvej 15
Institute for Theoretical Physics.

MF/C3 — Third Letter from Copenhagen (18th March 1933)

Copenhagen, 18.03.33

Dear Mother,

I received your letter dated the 14th. I am glad that you are all well. How is Maria's plan going? Urge them to study languages seriously as well, the knowledge of which becomes increasingly necessary for everybody.

Even here the weather is pleasant and temperate. I am not suffering due to the language. I read some Italian newspapers and speak in Italian with Placzek. The maids are sufficiently alert and have enough common sense for me to make myself understood.

There are two small Catholic churches in Copenhagen. The Italian community here is quite large. The majority are terrazzo workers, about one hundred and fifty, who cannot be replaced with local elements despite the high unemployment because of the inability of the Danes to learn the craft. There are also three or four Sicilians, not workers but prosperous merchants, and not just exporters of citrus fruits.

I'll go to some "Dante" [Society] meetings, which are attended by many Italians. The consul promised to invite me.

Bohr has left for about ten days. He is now in the mountains with Heisenberg to rest. He has been stubbornly contemplating the same problem for two years and the signs of fatigue have begun to show. In Copenhagen he is quite popular. The owner of a large brewery built, and offered him use of, a charming cottage that one accesses by passing through mountains of beer barrels. It is notoriously difficult to find for those going there for the first time. I went there once for tea. Bohr himself guided my steps, as I was fortunate enough to meet him as he was taking a leisurely bicycle ride around the area.

The people of Copenhagen are amazingly uniform, not only from physical and moral perspectives, but economic as well. Naturally, the devaluation of the krone causes some problems, making trips abroad especially expensive. The middle class is generally well-off without any major disparity in wealth, and used to be able to afford the luxury of traveling before the devaluation. This was imposed by the farmers to save the exportation of butter and lard to England, almost the only resource of the country. The commercial ties are connected with the spiritual orientation of the country, far more towards the UK than Germany. The press here generally expresses hostility to the latter. But these are superficial attitudes. The cultural isolation of Denmark is actually extraordinary. The diffusion of foreign books is mediocre, and they are rarely displayed in shop windows. The relationship with Norway is so lacking in intimacy that the once single language has now split.

I will write immediately to captain Fratto and wife. Best wishes and pass on my congratulations to Anita. Affectionate greetings and see you soon,

Ettore.

MF/C4 — Fourth Letter from Copenhagen (29th March 1933)

Copenhagen, 29.03.33

Dear Mother,

I received your letter confirming the good news given to me by father, especially regarding your health.

I do not require the *Nuovo Cimento* issues that I will find in Rome in a fortnight. With regards to Italian newspapers, *Corriere della Sera* is sufficient as it arrives here regularly after a day and a half.

For the last week we have had beautiful days here, sunny and warm. The temperature during the day reaches ten degrees. The high pressure system that prevails over the Baltic has produced beautiful weather here and cold currents and storms in Southern Europe.

Even here too much attention is paid to the international situation. It is believed that France will eventually join the bloc of the great powers, the only hope for peace for this country and its allies.

Bohr has returned. A new acquisition: Rosenfeld,[27] all that there is of theoretical physics from Belgium.

I'm still at the Berglund Hotel, which is good in itself and even better for its central location.

To learn German, I think the habitual reading of newspapers is very useful. You could buy one every day; perhaps you can find them in Via Po. The most common is the *Berliner Tageblatt*, which has become fascist out of the blue after editorial changes imposed by Hitler. I do not know whether it is good or bad that a German teacher speaks very little Italian; the important thing is that he or she knows the language and has an aptitude for teaching.

My health remains good. Copenhagen has all the amenities needed for material life. The inhabitants have almost ceased to interest me. Residing here for about a month has confirmed to me that there is not much to discover about the Danish soul. They are an extraordinarily

---

[27] What follows is Testimony *T/5*, written in 1964 by Léon Rosenfeld to Edoardo Amaldi.

peaceful, almost passionless people. It is unlikely that Denmark will ever make itself a talking point.

The most common means of transportation here is the bicycle. Almost all of the city's streets are lined by two bicycle lanes. There are relatively few cars, almost exclusively Fords. Petrol is cheap, 96 cents a liter. Apparently consumption taxes do not exist at all or are very low. Coffee is ten lira per kilo. Unemployment is high. Large numbers of unemployed are authorized to beg; an activity which assumes proportions here that are unknown elsewhere. They have no pretensions and thank you deeply for a donation of thirty cents.

It seems that the economic crisis should ease with time. In Germany, unemployment continues to decline even though the large industries have certainly not begun rehiring. Almost every country has by now achieved financial and economic stability. It is likely that the next economic conference will be able to work meaningfully towards ensuring a recovery. This will not involve a return to the pathological conditions of 1929, but the gradual emergence of new initiatives, and will therefore be slow but certain. I will leave around April 12.

Affectionate greetings,

Ettore.

T/5 — Testimony of Léon Rosenfeld (10th August 1964)

NORDITA
KØBENHAVN

Copenhagen, 10 August 1964

Dear Amaldi,

According to my promise, I am sending you my rather meager recollections of Majorana's visit to Copenhagen.
I am sorry that they amount to so little.
With kindest regards,

Yours sincerely
L. Rosenfeld.

Recollections about Majorana:

I remember Majorana turning up at the Institute in the company of Placzek. He was very shy and was content to listen to conversations without himself taking part in them, although one could see from the expression in his eyes that he was closely listening and had his own thoughts about what was being discussed. At that time, the usual language at the Institute, of course, was still German. Majorana certainly knew German very well, but he may have had difficulties in expressing himself in that language, and this may have been one reason for his silence. The only one, apparently, with whom he would talk freely was Placzek, presumably because he knew him from Rome. The result was that he clung to Placzek and was, in fact, never seen without him. Nevertheless, I did hear his voice once. We were together, the three of us, at a café near the

Institute (which in the Institute's slang was known as "Unter den Quanten"). Placzek and I conversed, and Majorana listened. At one time I expressed the wish to use some notes that Placzek had made. Placzek was willing to let me have them, but pointed out that they were rather long and that he had only one copy of them. I said: "Never mind, my wife will copy them." Then, suddenly, Majorana spoke and said, without moving a muscle in his face: "She's the ideal wife!" (L. Rosenfeld).

MF/L8 — Eighth Letter from Leipzig (6th May 1933)

Around April 12, Ettore returns home from Copenhagen for a brief visit in Rome. In early May, he travels back to Leipzig from Rome.

Leipzig, 06.05.1933-XI

Dear Mother,

I happily arrived at the scheduled time. My health recovered rapidly during the journey thanks to the Panflavin tablets. Remember this name: PANFLAVIN. You have to take at least one every hour, in severe cases, every half hour.

I had an excellent trip. It was a little crowded until Brenner. From Brenner to Monaco, I only had the company of a drunk Neapolitan gentleman. He was headed to Bremen to sell tomatoes and honored me with all his confidence since he presumed a priori that I was going to Leipzig for the same purpose.

The whole of Germany is now in springtime and there is no fear of a return of winter. It gives me great pleasure to see Leipzig again; it is far more likeable than it is ugly, and is said to be extraordinarily ugly. I have temporarily returned to my old hotel, and they are this very moment preparing the room I was assigned. I do not know if I will stay here for long. Perhaps I'll look for a room in the summer district of the city. This will depend on the companies that I find and the opinion of experts.

I will write to father at his Milan address as I heard at the station that he will be there in a couple of days.

I have to say goodbye because the preparation works around me are showing signs of slowing down and I would like to deal with the official correspondence.

Give my love to everyone,

Ettore

Institute for Theoretical Physics
Leipzig, Linnéstr. 5

D/CR4 — CNR: Internal Correspondence (4th May 1933)

NATIONAL RESEARCH COUNCIL
The Secretary-General

Rome, May 4, 1933-XI

Dear Bordoni,

Thank you for the brief note by Dr. Ettore Majorana, which after being converted into very a short letter to directors will be published in "La Ricerca Scientifica".

However, as it was work made possible by the scholarship procured by our CNR, it would have been preferable if Majorana had sent us a broad summary in Italian of his recollections of Zeitschrift. *The Italians who work abroad with our support have two obligations: to mention this in their memoirs that they publish abroad, and, when possible, to give our country the priority of any results, in Italian. Majorana still has time to prepare for publication in* "La Ricerca Scientifica", *a more extensive article on the subject he has worked on [...]*

Giovanni Magrini.

MB/L3 — Fourth Letter to CNR (9th May 1933)

Leipzig, 09.05.33-XI

Dear Professor,

I received your letter dated the 5th of this month; I will send an extensive summary of my work within two days. This has already appeared in the last issue of *Zeitschrift für Physik*: the journal was chosen because it was deemed opportune, given the connection with other work previously published by the same. I of course did not fail to highlight that my stay abroad was funded by the National Research Council.

I will use Italian magazines for any future publications, as per the desire expressed by the Direction.[28] I do not

[28] His boredom with these petty requests by the CNR bureaucrats of the time in comparison with the importance of work that Majorana was producing, is not *perhaps* alien to the fact that, in the end, the next article [the "already completed" one in German] never saw the light of day.

think however that we should avoid double publication when it is desirable to have our work immediately recognized abroad, given that the international diffusion of our physics journals, while reassuringly improving, remains very limited. If the case presents itself, I will request instructions on how I should proceed.

With sincerest greetings and many thanks,

Ettore Majorana.

MB/L4 — Fifth Letter to CNR (12th May 1933)

Leipzig, 05.12.33-XI

Dear Professor,

I enclose an almost complete version of all of my work on the theory of nuclei for inclusion in "Ricerca Scientifica" as per your kind request. I will send extracts of the German originals as soon as they come into my possession.

With cordial greetings and sincere thanks,

Ettore Majorana.

MF/L9 — Ninth Letter from Leipzig (15th May 1933)

Leipzig, 15.05.33

Linnéstr. 5, Institute for Theor. Physics

Dear Mother,

I had two letters from Rosina and father informing me of the ailments suffered by you last week. I hope that you have completely recovered and that with a strict diet and regular treatment in Karlsbad, you will heal completely. I have also received a letter from Turillo. I do not know if I can do anything for De Santis because I do not know anyone at the Faculty of Literature; I'll see if the consulate can give me some information. Even regarding the teacher Ottavi. Bordoni wrote to me; he wanted an article for the journal of the Research Council.[29] Some reprints from *Zeitschrift für Physik* should be arriving in the next few days. Please send twenty back to me as soon as you can.[30]

Leipzig, which was under a Social Democrats majority, accepted the revolution without any effort. Nationalist parades run frequently through the central and peripheral streets, in silence, but with a decidedly martial aspect. Brown uniforms are rare, but the swastika is everywhere to be seen.

The persecution of the Jews fills the Aryan majority with glee. The sheer number of those who will find work in public and private administrations following the expulsion of the Jews is very high, which explains the popularity of the anti-Semitic movement. In Berlin,

---

[29] Recall that Fermi, in urging Ettore to go to Leipzig to Heisenberg, arranged a six-month scholarship through the National Research Council (and for this purpose Ettore had written "Information on my didactic career", *Ms/1*: see Chapter 5).

[30] He is referring to the *Über die Kerntheorie* article on nuclear exchange forces, written in German and published in Germany (see Chapter 5).

more than fifty percent of the attorneys were Israelites. A third of these have been eliminated; the others remain as they were in office in 1914 and served in the war.

German nationalism resides largely in racial pride. All the school teachers have been advised to glorify the contribution to civilization of the Nordic race. In actual fact, it is not only the Jews, but also Communists and opponents of the regime in general which are being removed from social life in large numbers. Overall, the government is responding to a historical necessity: to give work to the new generation which is likely to be suffocated by the economic downturn.

It has been announced that Hitler will soon be visiting in Italy; I find no confirmation of this in Italian newspapers. The goodwill towards Italy is clear and profound. In contrast, the diffidence of the French elicits very lively irritation. There is no doubt that France is mistaken in underestimating the possibility of agreements with Germany, on the basis of concessions that would above all have a moral value for it. I await with great interest the statements that Hitler will soon make in the Reichstag, regarding the issue of armaments.

Give grandmother my wishes that she soon recovers from the mild indisposition which she attributes to a hereditary origin. Affectionate greetings,

Ettore.

MS/L1 — Letter to Emilio Segré from Leipzig[31] *(22nd May 1933)*

Leipzig, 22.05.33

Dear Segré,

Many thanks for your publication. This season's news: more credence is being given to Dirac's theory of positive electrons.

Heisenberg is treating it seriously. One of the most interesting consequences is that a quantum of sufficient energy can be absorbed by a force field, giving rise to a pair of electrons, one positive and the other negative. This may partially explain the absorption of heavy nuclei; Beck has performed calculations and found the right order of magnitude. The probability that any calculation by Beck need to be checked is considered small.

The internal political situation is entirely peaceful. The situation of the government can only strengthen with the improvement of international relations.

The issue of anti-Semitism must be judged in the context of the revolution that eliminated — as far as it could — all the opponents possible, among which,

---

[31] Initially believed lost, this much discussed letter was later published by E. Segré — 50 years after the disappearance of Majorana — in *Storia Contemporanea*, 19 (1988), p. 107. See Letter *T/16* by Segré, Chapter 11. For the contents of this letter, refer to the text (Chapter 6) and Letter *MG/L1* to G. Gentile, which follows.

almost without exception, the Jews. This does not mean that the Jewish issue in Germany is not in itself very grave; but the partial solution that was provided may have been affected by necessary political contingencies.

The Jewish situation in Germany is quite different to the one in Italy, both for the mood of the local Jews and for their number. The proportion of Jews in Germany may seem small at the light of the statistical lie (1%). In fact, they dominate finance, the press, political parties and in Berlin even represent numerical majorities in certain professions, as in the case of attorneys, for example. But neither religious nor racial prejudices alone suffice to explain the impossibility of a coexistence.

In Italy, we are accustomed to regarding the Jews as a historical survival which we respect, and do not take offence if some of them feel proud of their origins. Our policy, not of tolerance but of understanding, has been most fruitful and will continue to be so until the day, that cannot be far away, when the tradition of the jewish dealers approaches effortlessly that of the many Maritime Republics which honor the Italian people; one and indivisible.

In Germany the situation was quite different and, without analyzing the causes, it can be said with certainty that there was a Jewish question which showed no tendency of being resolved spontaneously. If the surgical maneuver could have been indeed replaced with the establishment of a policy, as firm as it was wise, that would have produced slower but more desirable results, only history will tell.

The fact is that what has led to anti-Semitism the almost unanimous suffrage of the Aryans is the existence of that silly and offensive thing that is Jewish nationalism. The German Jews in their majority were not europeanized, or in this case, germanized. It may be that this was due to the continuous inflow of fanatical elements from the eastern ghettos; at least this is the customary explanation given. But it is certain that the Jews asserted their separation from the Germans more or less with the same vigor as the latter, except for the ineffective last minute attempts at reconciliation, before the approaching storm. It is not conceivable for a nation of 65 million to be guided by a minority of 600 thousand who openly declared their desire to be a people unto themselves. Some say that the Jewish issue would not have existed if the Jews were familiar with the art of keeping their mouths closed.

But the current position of the Jews in Germany is not as serious as it may seem from the outside. The removal from public office was not total due to the known disposition in favor of the old ex-combatant employees. In certain categories almost two-thirds have retained their place. Remember that, under the empire, only baptized Jews were allowed to serve in public office. The vast majority of those who were engaged in private activities did not suffer from the change of environment, except for sporadic cases. The sentimental theory of the race is not credited with exaggeration and the moderate trend, which is content with the removal of the Jews from the management of public affairs, is fairly widespread.

Overall it is reasonable to consider the future of German Jews with a certain degree of optimism even if their fusion with the rest of the population will be delayed by the recent events. This may in any case indirectly have positive consequences if it serves to curb the dangerous Jewish immigration from the primitive communities of Slavic countries, especially from Poland. Among the newcomers to be singled out are the controversial rabbis who are accused of desiring this persecution to strengthen the unity of their people, which threatens to break up after the successful and peaceful coexistence with other peoples. An old story that repeats itself. But whatever developments the near future holds, we have to expect that in Germany, as in other countries still at issue with the Jews, after a more or less long march the civilization will not miss its aim.

I await Mussolini's speech, to which it is here attached decisive importance. I will read it tonight on tomorrow's newspapers. Give my regards to Fermi if he has not yet departed, with my best wishes for a safe trip.

Affectionate greetings,

Ettore Majorana.

MF/L10 — Tenth Letter from Leipzig (23rd May 1933)

Leipzig, 23.05.1933

Dear Mother,

I received your letter as well as a postcard signed by Padovani. Please return my most cordial greetings. I'm glad that you have recovered. The weather here is still cool and it must be the same in Karlsbad. The month of June, when you do not yet feel the excessive summer heat, is perhaps the most suitable for the treatment.

The German situation is very quiet; mine in particular even more so. I have a pleasant relationship with Heisenberg who enjoys my conversation, and patiently teaches me German. I must use this language exclusively after the departure of Bloch, who is well-versed in Tuscan.

The forthcoming departure of 24 Balbo airplanes for America is announced with the great fanfare. The role played by Germany in the preparation of the effort is exalted, both with respect to the choice of the route by von Gronan and for the organization of meteorological services, that it appears will largely be entrusted to German technicians. It is interesting to note the high-priced postal service that will be delivered by all the participating aircraft. Yesterday was the expected date of the much anticipated speech by Mussolini, who was supposed to announce the signing of the Four-Power Pact.

If you decide to leave soon for Karlsbad do not forget to bring, if you find them, some books that I must have forgotten: A guide of Germany, a small Leipzig guide and a de Agostini 1933 calendar which should contain a number of German stamps inside it.

Affectionate greetings.

Ettore.

MF/L11 — Eleventh Letter from Leipzig (2nd June 1933)

Leipzig, 02.06.1933

Dear Mother,

I received your letter dated the 29th of last month. As I have already told you, I have received the reprints and your other letter with the stamps. I'm pleased to hear of your travel arrangements to be implemented over the coming weeks. I'll be in Leipzig during the entire month of July. There is no hurry for books I left there. Ensure that the Austrian foreign currency control agent at Brenner notes on your passports the amounts of money you have with you exceeding four hundred lira per person. At the German border you need to go to the German customs office (inside the station at Kufstein, on the left) to obtain a statement of the amounts introduced. Even checks have to be reported.

The weather here remains quite good with mild and constant temperatures. Leipzig does not offer visitors an overwhelming number of wonders, but its friendly and hospitable inhabitants dissolve the many globally-diffused prejudices regarding Germans, one by one.

Public focus is directed mainly towards internal issues, especially the gigantic public works program that promises to provide jobs to five hundred thousand workers for two years and rehabilitate two million acres of currently unproductive land. They are making every effort to eliminate the importation of food, especially butter and substitutes that are of fundamental importance here. Of very high social importance is the recent government decision to provide loans on favorable terms to newlyweds. The economic situation is showing signs of improvement all around. The automotive industry is being privileged and has already had to increase the number of employed workers markedly. Regarding foreign affairs, the papers are deliberately silent on the Four-Powers Pact negotiations, so as not to generate over-enthusiasm, given that it cannot suddenly overcome the difficulties facing the review, especially, of the Polish border, the need for which is felt more strongly here. Relations with Austria seem to have become extremely tense. It is difficult to judge whether the majority of the Austrian people are really determined to maintain their independence. In any case, the Anschluss issue has lost relevance.

I learned with painful astonishment of the death of Venera who, for her age and vivacity, I thought was in perfect health. If you have any details regarding the cause of her death, please send them to me.

Greetings to all, including Lillì, and see you soon

Ettore.

MG/L1 — Fifth Letter to G. Gentile jr. (7th June 1933)

Physics Institute
Linnéstrasse, 5
Leipzig, 07.06.33

Dear Gentile,

I hope to see you again in Pisa while you prepare to conclude your first course of lectures. You'll forgive me if I begin with a recommendation. I know perfectly well that in these hard times the problem of finding work even for people of proven worth often presents insurmountable difficulties, but in this case I cannot turn to anyone but you for some hope of success. Mr. Werner Schultze of Hamburg was occupied until a year ago by Mondadori in a varied role which suited his artistic and literary inclinations. He was involved in the preparation of books, especially art, Italian, French, German, to translations, proofreading of texts, etc. Because of the economic crisis, he had to leave his position about a year ago and return to Hamburg. He would now like to settle in Italy, but even if there were some hope of being rehired by Mondadori, he does not think it wise to place undue reliance on this. He has thus turned to his friends (he befriended my family several years ago during one of his long stays in Rome) so they might, if possible, assist in his attempts to find suitable employment in Italy. I am aware that you know people in publishing circles, and if you think you can help by referring him to somebody influential, I shall gratefully consider it a personal favor to myself. If I receive a reply from you soon that is in some way encouraging (Mr. Schultze would also gladly accept a temporary position), I will inform him to be ready to provide, to those you identify, all the information that may constitute elements of judgment. I have no illusions about your ability to help me, but I'm sure that, if you can, you won't pass on the opportunity to credit to an old friend, who will not ignore the opportunity to prove his gratitude.

My work in the last month has been rather limited, also due to my poor health. At the physics institute it's life as usual. The somewhat toned-down seminars are dedicated to astrophysics, which [Friedrich] Hund has particular expertise in. Heisenberg is holding an elementary course on the theory of nuclei. The theory of [Paul A.M.] Dirac on positive electrons is being taken quite seriously. Heisenberg is studying the properties of relativistic invariance and the possibility of other applications, in addition to the calculation of the average life of positive electrons performed by Dirac. Of particular importance is the calculation, already attempted by

[Guido] Beck, of the probability of a light quantum of high energy to generate a pair of opposite electrons after colliding with a heavy nucleus. Beck has found orders of magnitude which suggest that to this supposed phenomenon might in part be attributed the observed deviation of heavy atoms from the [Felix] Klein and [Yoshio] Nishina formula.

I purchased Georg Webers' Universal History for four marks. We have little to rejoice in the treatment we are given there. Even though the last edition of this book was in 1929, when our pro-German policy was already in full development, there is no hesitation in qualifying Italy's entry into war as a "shameful betrayal" of her former allies. Of the heroic resistance of June on the Piave River, which really decided the fate of the war, there is scant note, and Vittorio Veneto is completely ignored. It seems that the Germans find it difficult to recover from their illness, which is to simultaneously cultivate the greatest possible number of enmities. The internal situation in Germany is apparently as stable as in Italy, but you cannot compare the political maturity of the two populations. Germany, which cannot find within its culture and history sufficient elements to found a common sentiment among German-speaking peoples, is forced to resort to foolish racial ideologies which are apparently not adequately echoed in Austria. The struggle against the Jews, though partly justified by instinct, is not justified by the reasons cited to support it; among these, sadly, the eternal issue of race dominates; and it is likely to soon end with little gained with respect to the sacrifices made, for the lack of clarity in the goals to be achieved. It seems that today the fate of the Four-Powers Pact will be decided. If Mussolini's happy initiative is successful, Europe will not only benefit from ten or twenty years of respite, but perhaps some of the contemporary pressing issues will gradually fade away or be resolved — when the bitterness left by the last war is forgotten, and the profound crisis surrounding falling birth rates since 1915 has changed not so much the numerical consistency as the features of some populations, particularly the Germans. In 1932: 978,000 births in Germany against 992,000 in Italy. In twenty years there will be as many young people in Germany as in Italy, but almost twice the number of old people. Then Germany will be conservative. Affectionate greetings,

Ettore Majorana.

MF/L12 — Twelfth Letter from Leipzig (8th June 1933)

Leipzig, 08.06.33

Dear Father,

I received your letter from Milan in which you announce your imminent departure for Rome. I await more details from mother regarding her journey. I am very pleased that she is still well; this will surely render her treatment in Karlsbad more effective and longer lasting. I will be receiving the visit of Rosina and Turillo with the greatest pleasure who, as mother writes, should take place towards the middle of the month. The season is perfect for traveling. Carefully plan the itineraries in relation to the available time; Germany is large and there are many cities worth visiting, even fleetingly. The costs of the trains are relatively low given the habit of traveling in third class. Even the general cost of living is not high, as you can easily imagine given the economic situation of the population. When the crisis is over and Germany is relieved of its debts (in the last three years, thirty-four billion liras plus interest have been paid) this will be the richest population in Europe. I do not think, however, that Germany can in the future constitute a danger to peace. Of course many Germans dream of revenge, as would any other people in their place, as did the French with much less reason from 1871 to 1914. But before the military map of Europe is affected by such changes to render new adventures possible, the nature of the German people will already have changed as a result of the tremendous crisis of falling birth rates which began in 1915. In thirty years, the Germans will be the oldest and consequently the most peaceful people in Europe. The Austrians, on which the myth of race does not seem to exercise much influence are already far ahead on this path. There is some probability that Austria may actually become like a second Switzerland as desired by

France. I read the text of the Mussolini pact; wonderful for the frugality of mutual commitments; but it is the indispensable introduction for the consolidation of peace in that it envisages collaboration between the four powers that be enduring rather than sporadic.

Affectionate greetings,

Ettore.

MF/L13 — Thirteenth Letter from Leipzig (14th June 1933)

Leipzig, 14.06.1933

Dear Mother,

I received your letter dated the 9th of the current month informing me of your new travel plans. Paris is an excellent choice, and it is good that you are taking advantage of the opportunity by not making your visit too short. I've also heard about your pleasant trip to Florence for the May music festival. The performances held in the Boboli Gardens were widely reported in the German newspapers. I hope that Luccio enjoyed his trip in northern Italy. Doesn't he have a passport yet? I believe he can obtain one now quite easily, while he will need to wait a long time when the applications are numerous again.

I am going to buy linen thread, even if there is no real hurry. The weather has taken a turn for the worse here and the sky is pitch black today. But it is unlikely to last.

The rest of this letter has been lost.

MF/L14 — Fourteenth Letter from Leipzig (23rd June 1933)

Leipzig, 23.06.33

Dear Mother,

I received your letter, together with father's. I believe mine will arrive in time before your departure. I am very pleased that the program you are following has already had such beneficial effect on your health. You must stick to it until you are completely cured.

The weather here is still as bad as in most of Europe, but given the late season, it will likely recover quickly. You can assume summer has really begun, possibly to set the exact date of your departure, when the forecast bureau signals the end of the low pressure system that has persisted throughout Europe for several weeks.

I will not need any money until July 10, but it is preferable, even if for that date you plan to be in Leipzig, that you have the checks issued directly in my name at the Central Bank in Rome, in order to facilitate cashing them. Even for your travel expenses you would do well to use checks as these are less cumbersome than notes and are paid more.

If you go to Paris before going to Karlsbad, it might be interesting to break up the trip with a quick stop in Southern France: in Nice, or possibly in Marseille: the Naples of France where one hundred and fifty thousand Italians live. Northern France has significantly different characteristics to Southern France and the opportunity, starting from the closer region, to experience both should not be missed. I believe that you will be well-received wherever you go: the French are very open to a renewed friendship with Italy, and the longer it is desired, the more it is appreciated.

I hope to soon receive your good news confirming the date of departure and subsequent news or impressions regarding your trip to France. I would like to know if father will be staying in Rome for a long time.

With warm greetings and best wishes for a good trip,

Ettore.

MF/L15 — Fifteenth Letter from Leipzig (1st July 1933)

Leipzig, 01.07.33

Dear Mother,

I received the registered mail from Rome and Maria's postcard from Paris. I am pleased you journeyed well and I hope that you enjoy yourselves very much. Do not leave out anything indicated as worth seeing in a good guidebook. Above all, the art collections which Maria and Rosina are especially interested in.

When you have decided, let me know the exact day of your arrival. If you want to stay in a hotel, mine is decent and I do not think there are any better nor worse in Leipzig. But it is certainly preferable for you to eat in restaurants so you can choose what you eat. There are some excellent and inexpensive restaurants.

Here, the weather seems to be giving the first signs of improvement; I hope that it has been better in Paris. You did not mention if father is still in Rome and if he will remain there for a long period. I await your answer so I can write to him. I'm sure that when you arrive you will have many things to tell about your stay in Paris, but I'd really appreciate any details you anticipate by mail, especially regarding Maria. Be sure to encourage Maria, Rosina and Turillo to collaborate in the creation of a detailed program of visits, excursions, etc., in order to make the most of all the available time.

With many affectionate greetings and best wishes and hope to see you again soon,

Ettore.

MF/L16 — Sixteenth Letter from Leipzig (25th July 1933)

After the visit of his family in July 1933 in Leipzig, Ettore resumes addressing his letters to his mother in Rome (the previous letter is addressed to Paris):

Leipzig, 25.07.33

Dear Mother,

I received your second postcard. I am surprised that you have not received my letter sent on the 18th of this month.

There was also another letter attached from Nitto with trifles and payment information. I did not read it carefully and I cannot repeat the details. I am still under the doctor's care, with slow, but in his opinion, sure results.

I have no intention of coming to Abbazia because I would not be able to swim and the heat at the beach would be unbearable. I would like instead to go directly to Rome for a few weeks. If there are difficulties with regard to, laundry service, sending keys, etc.., please list them and I will consider how to resolve them. I have had some extra expenses and the money that I have will not suffice for the trip.

Father wrote to me from Milan. He is dealing with delicate mechanics experiments. He had a telephone conversation with Rosina shortly after his arrival.

I have asked the famous Argus to gather information in Hamburg. I do not know what the best route for Abbazia is. With warm greetings and best wishes for a good trip,

Ettore.

MF/L17 — Seventeenth Letter from Leipzig (27th July 1933)

Leipzig, 27.07.33

Dear Mother,

I received your letter, and then one from Turillo. Your worries concerning my intention to go directly to Rome seem to me exaggerated. My health is already showing signs of marked improvement and I no longer have serious problems unless I cease to follow the doctor's prescription. I only need a little patience and a little time to heal, perhaps not more than a couple of weeks. I think that I will feel better in Rome than in Abbazia. Also for meals, because no matter how great the skill of Mr. Ball may be in satisfying the changing needs of my stomach, I think that I could still not expect from him the variety in any Roman restaurant. I am no longer on a milk diet; I must indeed mainly eat vegetables. I do not use any butter. The ordinary choices in Rome are beyond all comparison and are unconditionally hailed by all foreign physicists with stomach problems who have had the pleasure of experiencing them. Regarding the heat, I know what it is like as I have spent August in Rome before without having accumulated any disdainful recollections. As for the service, in the worst case I will not have any difficulty in doing everything entirely by myself with the necessary diligence. I do not understand why you say you want to come to Rome for a few days; the conscientious Borromeo will be just as conscientious even in your absence, and I will ask for help if I need it. You would bring me useless sorrow if you embark on so long and tiring a journey with no purpose or justification. But I do not intend to change my plans out of fear that you will carry out so unreasonable a threat.

I think you can send me the check from Abbazia. I will depart shortly after I have received it.

Affectionate greetings,

Ettore.

MF/L18 — Eighteenth Letter from Leipzig (3rd August 1933)

Leipzig, 03.08.1933

Dear Mother,

I received your letter and the check. I'll leave tomorrow evening for Rome, where I shall arrive on Sunday morning. My condition continues to improve. The Italian *avanguardisti* [members of the Italian fascist youth organization] passed through Leipzig and have been triumphally received throughout Germany. Some of the highest political authorities have prepared speeches for them, expressing the general desire to raise Germany to the same degree of civilization attained by Italy. The papers discuss the martial aspect of the *Balilla* and *Avanguardisti*, and the many officers who accompany them, all gymnastics masters from the 'Foro Mussolini' sports complex.

Give my regards to old acquaintances, if there are any: Ms. Aurora and Pina, the Pirate, the Captain, Padovani, and so on.

Affectionate greetings,

Ettore.

MB/LS — Sixth Letter to CNR (4th August 1933)

Leipzig, 04.08.33-XI

Dear Professor,

As of July 31, my summer course in Germany has concluded as well as the period established for my residence in this city. I will therefore return to Italy and proceed to provide a report of my current work when it is completed.

With cordial greetings,

Ettore Majorana.

MF/RS — Letter From Rome to His Mother (15th August 1933)

Having returned home from Leipzig, Ettore now writes to his mother from Rome. This letter, like the previous one, is addressed to Abbazia.

Rome, 15.08.33

Dear Mother,

I received your letter. Mrs. Moratti paid 800 lire, which in her opinion and according to the recollections of Annita is the amount of the monthly rent. The receipt that you left me was for 850 lire and I had to correct it. Let me know if it's okay.

Grandmother has no specific complaints but is still weak and suffering from the heat. I will have her visited by the doctor in a few days. I'm feeling a bit better.

Great excitement for the arrival of Balbo. I listened to the celebrations on the radio.

I received the information you wanted. I did not find anything very important and I do not think that you are in a hurry to know the details.

Affectionate greetings,

Ettore.

MQ/R4 — Letter to Quirino Majorana[32] *(6th September 1933)*

Rome, 06.09.1933

Dear Uncle,

I received your postcard. I have been in Rome for a month. Even the rest of the family has arrived one or two at a time, and will perhaps travel to Monteporzio in a while. I will certainly stay in Rome and you can write me preferably at home because I rarely go to the institute.

---

[32] Mrs. Silvia Majorana in Toniolo, daughter of Prof. Quirino [illustrious and accomplished experimental physicist, former president of the Italian Society of Physics] entrusted Franco Bassani and me with copies of 34 letters written by Ettore to his uncle Quirino between 17th March 1931 and 16th November 1937. Here we publish only a few, as they are predominately technical-scientific in nature. Two of them from 1937 (the last two given here) were reproduced by Bruno Preziosi and S.I.F. in the volume published by Bibliopolis (Naples) in 1987, cited elsewhere. Subsequently, they have been all published by a Colleague (in 2008).

I will not return to Leipzig this year. I believe father has interrupted his research. I only found Rasetti in Rome; Fermi will return from his trip to America soon.

Affectionate greetings,

Ettore.

MB/R2 — Seventh Letter to CNR (14th September 1933)

Rome, 14.09.33-XI

Dear Professor,

In answer to your kind request [...] I have the pleasure of informing you of the following. My stay abroad took place over two periods [...]

I published a work "On the theory of nuclei" in *Zeitschrift für Physik*, the leading German physics journal. It contains important corrections to the theory developed by Heisenberg in three previous studies. Prof. Heisenberg has accepted the findings in my work and has given it wide exposure in a recent lecture course held in Leipzig, and in a general report on nuclear physics intended for an international conference (currently being printed). In the last period of my residence in Leipzig I began other projects that for health reasons I could neither complete nor bring close to completion. I think it useless to talk about these. I enclose, for your convenience, another reprint of the first paper.

With cordial greetings (also from my father) and with renewed thanks for your constant interest,

Ettore Majorana

Viale Regina Margherita 37 – Rome.

MG/R4 — Sixth Letter to G. Gentile jr. (27th July 1934)

Monteporzio Catone (Rome), 27.07.34

Dear Gentile,

Thank you for sending the book by Jeans in your beautiful edition (and translation)[33] Which I am now well into as it fills my rural idleness. I very much liked the profound preface, truly suited to the Italian public for the correct references to streams of thought which are here dominating. I believe that the greatest merit of this book is to anticipate the psychological reactions that the recent development of physics will inevitably produce when it is generally understood that science has ceased to be a justification for vulgar materialism. I therefore believe that your translation will seriously help rekindle an Italian interest in scientific problems.

I will pass by Rome in a few days and I very much hope to have the opportunity of seeing you again.

Yours sincerely,

Ettore Majorana.

---

[33] James Jeans, *I nuovi orizzonti della scienza [The New Background of Science]* (Sansoni, Florence, 1934); translated into Italian by G. Gentile jr.

MQ/R35 — Letter to Quirino (20th February 1935)

Rome, 20.2.1935

Dear Uncle,

I received your letter in which you announce new experiments concerning the action of light on the resistance of thin metallic foils. It is very important that you were able to separate the thermal effect from the supposed photoelectric effect. You do not however describe… [*omissis*]

Thank you for wishing me well with my activities, which I hope will soon return to their normal levels (which are not a lot). With affection from everyone to you and your family

Ettore.

MQ/A35 — Letter to Quirino (11th August 1935)

Abbazia, 11.08.35

Dear Uncle,

I received your last letter. Regarding the experiments… [*omissis*]

It is not necessary that on page 8 you give me credit for an elementary calculation, which you can have recalculated or checked by anyone you see fit if you do not want take the responsibility. I'm leaving today for a slightly uncertain destination; I'll send you my new address as soon as I have achieved relative stability.

Affectionate greetings,

Ettore.

MQ/R13 — Letter to Quirino (5th September 1935)

Rome, 05.09.35

Dear Uncle,

I received your last letter. In my previous letter I simply wanted to point out the difficulty of giving a simple yet comprehensive explanation of all the observed facts. I did not at all intend to exclude the possibility that light can produce extra-thermal variations of the resistance, but, as it seems even you have admitted, we run into serious unlikelihoods in the interpretation of the "tail" phenomena. Of course, as you strongly reaffirm that there is a very low probability of any experimental perturbations, you can overlook the theoretical improbability on the basis of the evidence of the observations. I do not understand why, however, when faced with two conflicting improbabilities, you should definitively accept one of them when it would be so simple to resolve the dilemma with further research.

I do not think that you would simply respond that an ascertained fact cannot be denied by theories, and especially by theoretical assumptions. What we have is not actually immediate evidence, but facts, recognized with the help of certainly obvious hypotheses on the experimental conditions, but which, without being overly subjective, cannot be considered as having a greater likelihood than the theoretical presumptions. For these reasons, it would seem good practice to also take account of eventualities, that in other circumstances would be negligible, being a priori highly improbable. I therefore find your accusation of not being in a serene state unjustified; I have indeed tried not to attach to my objections (that I was forced to raise) excessive value and even less decisiveness. Of course, if you refer to this with respect to the practical aims of life, no one can presume that you continue with experiments that may present more and more new problems, and if you want to leave this task to others I would certainly not forward any criticism. But I did feel obliged to point out which are, in my humble opinion, the unresolved problems at the present time. I would be very sorry if there remained the slightest trace of any misunderstanding.

Affectionate greetings,

Ettore.

MQ/R14 — Letter to Quirino (16th January 1936)

Rome, 16.01.36

Dear Uncle,

I received your last letter. I learned from Gentile of your presence in Rome. I look forward to viewing the results of the research that you say are currently in progress and I wish it the greatest success. I have been dealing with quantum electrodynamics for some time.[34] My health è is now sufficiently good; thank you for your interest in the matter. Affectionate greetings, also to Aunt Vincenzina, to Carmela and to Silvia, who I know better by reputation than personally.

Ettore.

MQ/R18 — Letter to Quirino (23rd August 1936)

Rome, 23 August 1936

Dear Uncle,

I have received your express letter. It is clear that your interpretation is hypothetically admissible. The geometric representation of currents... [*omissis*]

There is no reason to include citations. One can always in fact presume that the formulas relating to thermal conduction were taken from dusty old tomes. It is therefore also a good idea to write down the mathematical solution, omitting interim developments that are always inspired by well-known models.

Affectionate greetings,

Ettore

---

[34] Considering Ettore's modesty when expressing himself, this very importantly means that, during 1935, Majorana had been intensely involved in new research in the field of *quantum electrodynamics*. Probably the notes with the new findings that no doubt resulted, are to this day still missing.

P.S.: Luciano is here, quite busy, and I am taking advantage of this because I do not like the sea air. Greetings, Ettore.

MG/R5 — Seventh Letter to G. Gentile jr. (20th June 1937)

Monteporzio Catone, 20.6.1937

Dear Gentile,

I sincerely thank you for your lovely book *Fisica Nucleare.*[35] It is truly a perfect example of the informational genre; and the amount and variety of smoothly displayed data makes it enjoyable and extremely interesting for anyone, even those with limited technical knowledge. I hope your publisher knows how to "launch", because nothing like this has been seen in Italy for long time, nor will there be any soon. It should really pass through the hands of everyone. Affectionate greetings,

Ettore Majorana.

MG/R6 — Eighth Letter to G. Gentile jr. (25th August 1937)

Monteporzio, 25.8.1937

Dear Gentile,

Thank you for your letter and for your study on polarizing slits that I received some time ago. Even though I am

[35] G. Gentile jr., *Fisica Nucleare* (Ed. Roma Pub., Rome, 1937).

not familiar with the topic, I was able to see how your preparation is solid and sophisticated even in this field of classical physics.

As you must have already guessed, I'm still in Monteporzio, and I also look at the sky (the sea from a long distance) and I can see the daily failures of the weather forecasts. I can moreover cultivate astronomy.

I think your outright diffidence regarding Fermi is unjustified; he spoke of you with the most sincere sympathy. As for the other members of the committee, either I have never seen them, or I have not seen them for a very long time. But it seems at least one of them should have the authority and the will and the duty to vouch for Giovanni Gentile.

I haven't heard anything about the work that I published, in N.C.,[36] apart from a Swiss man asking me for a reprint. I fear he will spoil his holiday. But I am not on holiday; all the more so as, since the evening of 15 August, the weather has cooled and staying in the hills is beginning to become pleasant. I estimate that I will be in Rome in two and a half weeks, but only for a few days. If you are there, we should make ourselves available to see each other.

Affectionate greetings,

E. Majorana.

[36] This refers to the article "Teoria simmetrica dell'elettrone e del positrone", *Nuovo Cimento*, 14 (1937), p. 171, now quite famous.

MQ/R28 — Letter to Quirino (1st September 1937)

Monteporzio Catone, 01.09.1937

Dear Uncle,

I have received your letter. I am truly pleased that you have now entered the period of convalescence [...] I'm just sorry that I have been quite ill and have not be able to come and see you during your stay in the clinic [...]

Dalla Noce informed me of the congress dedicated to Galvani, which I'd like to attend but I may have to miss to avoid traveling too far from Sicily. I improvised an opening talk according to what it seems that you wanted. You might judge it to be too generalized, but I found it difficult to delve much further without exceeding the specified limits. In any case, if you do not like it, send it back to me with your comments [...] Regarding the talk, the opening and closure are not, by circumstance, models of eloquence, but I added them for their possible connections with the rest [...]

Yours,
Ettore.

MQ/R29 — Letter to Quirino (16th November 1937)

Rome, 16 Nov. 1937-XVI

Dear Uncle,

Thank you very much for your letter. I gave Sgroi the necessary instructions. I also received the "Sapere" [magazine]. Your talk has been much admired, and by many.

I laughed a little at the procedural oddities surrounding my competition, of which I had no idea. I truly hope to go to Naples.

Rosita is here for a brief stay; the others are in Sicily.

Many affectionate greetings, also to Aunt,

Ettore.

MG/R7 — Ninth Letter to G. Gentile jr. (21st November 1937)

Via Reg. Margherita 37

Rome, 21 Nov. 1937-XVI

Dear Gentile,

I received your wonderful, long letter. I thought you would already be set up in Milan by now, and I continue to believe that you will not have to wait more than a few days. It is true that I foresaw a slightly different trio, but I knew Wick had to be the first.

I have seen the new work of Racah, but only in proof. In the second part there is something real: that is, the actual application to the ß theory and the criticism aimed at me. The first part is not original and even the mathematics is shaky: Racah does not know, or does not believe, that *spinors* have two values, and neglects the consequences. Things that always occur when you learn from others (Pauli) rather than by yourself.

I do not know if and when I will go to Naples. I am in correspondence with Carrelli, who is really a very nice person (his maxim: men are much better than you would believe). Even Segré and all the others have been very kind. I marvel that you doubt the strength of my stomach, in the metaphorical sense. Pius XI is very old and I received an excellent Christian education; if at the next conclave they make me Pope for exceptional merits, I will without doubt accept. Excuse me if I stop at the first half kilometer.

Many affectionate greetings and best wishes,

E. Majorana.

MN/1 — Letter to Angelo Savini-Nicci (30th November 1937)

Rome, 30 Nov. 1937-XVI

Dear Angelo,

I am very moved by your well wishes from Addis Ababa, which are indeed a nice touch. I am glad that you still remember me. I send you in return my best wishes for your fervent life as a pioneer and the bright future that awaits you.

Affectionately,

Ettore.

MF/N1 — First Letter from Naples (11th January 1938)

Ettore starts writing home again after almost five years, when he leaves Rome to take up the position as professor at the Royal University of Naples.

Naples, 11.01.1938

Dear Mother,

I have announced the beginning of my course for Thursday the 13th at nine.[37] But it was not possible to verify that there were no time conflicts, so it is possible that students don't come and it will have to be postponed. I met with the dean, with whom I have agreed to avoid any official nature to the opening of the course, and for this reason I would advise you not to come.[38] Carrelli has been very nice to me and today we bought furniture for my room, courtesy of the Faculty. Effectively, the Institute is reduced to Carrelli himself, the senior assistant Maione and the young assistant Cennamo. There is also a professor of earth's physics who is difficult to discover. I found a two-month-old letter from the Rector which announces my appointment

---

[37] In what follows are Ettore's notes for his inaugural lecture. Found in 1972, we published them in 1982 (*Corriere della Sera* on 19.10.82) to mark 50th anniversary of much of his most important research.

[38] The family, however, promptly arrived at nine o'clock on Thursday, 13th January 1938, to attend the inaugural lecture of Ettore: as his sister Maria recalled.

"for high fame of singular expertise". Not finding him, I wrote him a letter with an equally elevated response. Carrelli prepares mechanics lessons with many plays. The dominant concern is for the exercises, at least for Carrelli and assistants. The institute is very clean and tidy, although it has little equipment.

Hotel "Napoli" is decent, with reasonable prices; so it is likely that I will stay there for a while. Naples, at least in the central part, has a very pleasurable aspect, although it is strange for the lack of vehicles. I will write to you on Thursday about the first lesson.

Affectionate greetings,

Ettore.

Ms/2 — Notes for the Inaugural Lecture for His Course (13th January 1938)

In this first introductory lesson, I will briefly discuss the aims of modern physics and the meaning behind its methods, above all in as much as they contain more of the unexpected and original, compared to classical physics.

Atomic physics, which will be our main topic, remains primarily a science of enormous *speculative* interest due to the complexity of its research, which reaches to the deepest roots of natural phenomena, in spite of its numerous and important practical applications, and the wider, perhaps revolutionary significance that its future may have for us. I will therefore firstly discuss, without any reference to special categories of experimental facts and without the help of the mathematical formalism, the general features of the concept of nature accepted in new physics.

* * *

A) The *classical physics* (of Galileo and Newton) at the beginning of this century was entirely connected, as we know, with the *mechanistic* concept of nature that from physics spread not only to the related sciences, but also to biology and even social sciences, characterizing not so long ago all the scientific and a good part of philosophical thought; even if, to tell the truth, the usefulness of the mathematical method which was its the only valid justification always remained exclusively confined to physics.

This conception of nature was essentially based on two pillars: the objective and independent existence of matter, and physical determinism. In both cases they are, as we shall see, notions derived from common experience and then generalized and rendered universal and infallible, especially for the irresistible charm that, even on the deepest souls across all time, has been exercised by the exact laws of physics, considered the true sign of the absolute and the revelation of the essence of the universe: the secrets of which, as Galileo stated, are written in mathematical characters.

The *objectivity* of matter is, as I said, based on a notion of common experience, because it explains that material objects have an existence in themselves, regardless of whether or not they are subject to our observation. Classical mathematical physics added to this elementary observation the detail or the assumption that this objective world can be mentally represented in full accordance with reality, and that this mental representation may consist in the knowledge of a set of numerical magnitudes sufficient to determine at each point in space and at any time the state of the physical universe.

Determinism is instead only partly a notion of common experience. This actually gives contradictory indications on the subject. Beside the facts that are inevitable, like the fall of a stone dropped in a vacuum, there are others — and not only in the biological world — in which the inevitability of the sequence is less evident. Determinism as a universal principle of science could therefore only be applied as the generalization of the laws that govern celestial mechanics. It is common knowledge that a *system* of points — which, in relation to their enormous distances, can be considered the bodies of our planetary system — moves and changes according to the laws of Newton … [*omissis*] … It follows that the future configuration of the *system* can be foretold through calculation, provided that you know the initial state (that is, the set of positions and velocities of the points comprising it). And everyone knows of the extreme rigor of astronomical observations that have confirmed the accuracy of Newton's law; and how astronomers are actually able to predict with this law alone, and over great distances in time, the precise minute when an eclipse, or an alignment of planets or other celestial event will take place.

* * *

B) To illustrate *quantum mechanics* in its current state, there are two almost opposite methods. The first is the so-called historical method: and it explains how, from precise and almost immediate experimental indications, the first idea of the new formalism came into being; and how this then developed in a way that was far more compelled by internal necessity than by heeding new decisive experimental facts. The other method is the mathematical one, according to which quantum formalism is immediately presented in its more general and therefore clearer form, and only after are the application criteria illustrated. Both of these methods, if applied separately, have very serious drawbacks.

It is a fact that, when quantum mechanics was born, it was for some time met with surprise, skepticism and even complete incomprehension by many physicists. This was mainly due to the fact that its logical consistency, internal coherence and sufficiency appeared, more than dubious, elusive. This was also, though completely erroneously, attributed to a specific obscurity of exposition by the initial founders of the new mechanics; but the truth is that they were physicists, and not mathematicians, and for them the evidence and justification of the theory essentially consisted in the immediate applicability to the experimental facts that had suggested it. The general formulation, clear and rigorous, came later and partly due to mathematical minds. If we therefore simply repeated the exposure of the theory according to its historical aspect, we would immediately create a useless sense of discomfort or distrust feeling that were justifiable in the past but which can no longer be accepted, and can be spared. Furthermore, the physicists — who have managed to clarify, not without difficulty, the quantum methods through conceptual experiences imposed by their historical development — have almost always at some point felt the need for greater logical coordination and a more perfect formulation of the principles, and have not refused the help of mathematicians for this task.

The second method, the purely mathematical one, presents even greater inconveniences. It in no way permits an understanding of the genesis of the formalism and consequently the place that quantum mechanics has in the history of science. But above all it completely fails to fulfill the desire to somehow intuitively perceive its physical significance, often so easily satisfied by classical theories. Its applications, then, while uncountable, appear rare, disconnected, and even modest considering its overwhelming and incomprehensible generality.

The only way to make it easier for those who are now beginning to study atomic physics, without subtracting anything from the historical origin of the ideas and the very language we use today, is to start with the broadest and clearest discussion possible of the mathematical tools that are essential to quantum mechanics, so that the student will already be familiar with such instruments when the time comes to implement them, and will not be frightened or surprised by their novelty: one can thus rapidly proceed to derive the theory from experimental data.

These mathematical instruments already generally existed before the onset of new mechanics (as disinterested work by mathematicians who did not realize such exceptionally wide fields of application); but quantum mechanics "forced" and extended them to satisfy practical needs; thus we will discover them, not with the criteria of mathematicians, but with those of physicists. That is, without worrying about excessive strict formality, which is not always easy to achieve and often absolutely impossible.

Our only ambition will be to discuss as clearly as possible the effective use physicists have made of these tools for over a decade; the use of which, having never created any difficulty or ambiguity, constitutes the essential source of their certainty.

Ettore Majorana.[39]

MM/N — Reply to the Minister of National Education (12th January 1938)

Institute of Experimental Physics
of the Royal University
of Naples, 12.01.1938-XVI

Subject: Gratitude
to H.E. the Minister for
participation of appointment

To the Directorate General of Higher Education

I receive direct communication from His Excellency the Minister for my appointment as Professor of Theoretical Physics at the Royal University of Naples, pursuant to art. 8 of Decree n. 1071, 20 June 1935-XIII. Respectfully conveying to His Excellency the Minister my heartfelt expression of gratitude for the high honor bestowed unto me, I would like to assure that I will give all my energy to the school and to Italian science, today so fortunately rising to regain their ancient primacy.

Respectfully yours,

Ettore Majorana.

[39] This manuscript, *Ms/2*, consisting in notes drafted for personal use and not for publication, has been very slightly "edited".

MF/N2 — Second Letter from Naples (22nd January 1938)

INSTITUTE OF EXPERIMENTAL PHYSICS
Royal University of Naples
Via A. Tari, 3

22 January 1938-XVI

Dear Mother,

I received your letter and parcel of linen. I do not have a cold. I have now finished the fifth lesson. I'm still at [*the hotel*] Terminus, but I'll soon move to a residence. Carrelli is still in Rome. The weather is good again. I will be short of money next week; therefore, could you please ask Luciano to withdraw my part of the account from the bank and perhaps send it all to me, taking into account previous withdrawals and after repaying you the thousand lira that you have gave me recently. The nurse provided me with some good addresses for residences. I think I'll come in a few days but only for a few hours because I have to pick up a book by Treves and others from home.

Affectionate greetings and see you soon,

Ettore.

MF/N3 — Third Letter from Naples (23rd February 1938)

Naples, 23.02.1938-XVI

Dear Mother,

I am in Hotel Bologna, via Depretis, which is quite good and very clean. The staff is almost all from Bologna. I have a decent room; today they will give me a better one on Via Depretis, from which in three months' time I will be able to see Hitler pass by. Have you all overcome your minor colds? I may come after Carnival.

Affectionate greetings,

Ettore.

MG/N1 — Tenth Letter to G. Gentile jr. (2nd March 1938)

INSTITUTE OF EXPERIMENTAL PHYSICS
Royal University of Naples
Via A. Tari, 3

02.03.1938-XVI

Dear Giovannino,

I received the postcard you addressed to Rome. I have spent the Carnival period here; you can easily imagine the follies. All of Naples is in preparation for the upcoming visit of Hitler. My classes will resume on Saturday. I'm pleased with my students, some of whom seem determined to take physics seriously. I hope to see you again soon.

Affectionate greetings,

Ettore Majorana.

MF/N4 — Fourth Letter from Naples (9th March 1938)

Naples, 09.03.1938-XVI

Dear Mother,

I received your letter. I will take all necessary precautions with the linen. Some signs boast with great emphasis that the hotel laundry and ironing services are impeccable. The weather is beautiful here, perfect for sailing in the gulf. How are the winter sports going? I believe that Maria has already made great progress in painting and she can soon send me a photograph of her most famous works … I hope to come at the end of the week.

Affectionate greetings,

Ettore.

MF/N5 — Fifth Letter from Naples (19th March 1938)

Naples, 19.03.1938-XVI

Dear Turillo,

I received your letter. I cannot come at the moment because on Monday I have to deal with some things at the registry office and elsewhere.

I'll see if it is possible to obtain the booklet[40] for Mother, but I do not see how we can say we live

[40] The "social security booklet" for State health assistance.

together because I have am obliged to take up residence in Naples, in fact I have already done so temporarily here in the hotel, alias Via Depretis 72.

I am sending you a telegram so you do not wait for me to arrive tonight, but I will certainly come next Saturday.

Affectionate greetings,

Ettore.

MC/N — Letter to Carrelli from Naples (25th March 1938)

Naples, March 25, 1938-XVI

Dear Carrelli,

I've made a decision that is by now inevitable. There is not within it a single grain of selfishness, but I recognize the bother that my sudden disappearance will bring to you and the students I ask your forgiveness for this also, but above all for having betrayed the trust, sincere friendship, and kindness that you have demonstrated to me over these months. I ask you also to give my best to all who I came to know and respect at your institute, Sciuti in particular, for all of whom I will hold dear memories, at least until eleven o'clock this evening, and possibly thereafter.

E. Majorana.

MF/N6 — Letter "To My Family" (25th March 1938)

This was left in his hotel room in an envelope addressed "To My Family":

Naples, March 25, 1938-XVI

I have just one wish: that you do not wear black. If you wish to mourn me then do so, but not for more than three days. Afterwards, if you can, keep my memory in your hearts, and forgive me.

Affectionately,

Ettore.

MC/P — Letter to Carrelli from Palermo (26th March 1938)

Sent by express mail on the morning of Saturday 26th from Palermo (after an urgent telegram) and arrived at its destination in Naples, the next day, Sunday morning (see Chapter 1).

GRAND HOTEL SOLE
PALERMO

Palermo, March 26, 1938-XVI

Dear Carrelli,

I hope that the telegram and the letter arrived together. The sea has refused me and tomorrow I will return to the Bologna Hotel [in Naples], perhaps together with this letter. I do, however, intend to renounce teaching.

Do not take me for an "ibsenian girl" because the case is different. I am at your disposal for additional details.

Affectionately,

E. Majorana.

T/6 — Testimony of Vittorio Strazzeri (31st May 1938)

Dear Mr. Majorana,

It is my firm belief that, if the person who traveled with me was your brother, he did not kill himself, at least not before our arrival in Naples.

This is because when I arose [from the bed], *we were in front of the port of Naples and many passengers were on the deck of the ship as it was a* very clear day.

I repeat that I did not see any luggage in the compartment, but what caught my attention was that his vest, or perhaps it was a jacket (in any case, some clothing), had been laid on the netting that covers each bed; this caught my attention because my biggest concern when traveling is the safekeeping of my wallet and passport.

I do not question that the third passenger was called Carlo Price, *but I can assure you that he spoke Italian like* we from the South do, *and also that he seemed to me to be some kind of shopkeeper, or of lower status, as he lacked that unconscious refinement which comes from higher culture…*

I repeat: if the young man who traveled with me was your brother (I say "young" because his hair was full, and because he gave me that general impression), he certainly did not kill himself before the ship's arrival in Naples.

Please kiss your mother's hands for me, and give my regards to your family.

If you have any news, please inform me, and believe me that — if it is good news, as I hope and believe — it will bring me great joy. With devotion.

Yours
V. Strazzeri
Palermo, 31.5.1938

P.S.: Forgive me if I dare to offer you a suggestion, which is to explore the possibility that your brother has closed himself up in a convent, as has happened even with people who were not very religious, I believe at Monte Cassino.

T/7 — Letter from Ettore's mother to Mussolini (27th July 1938)

[Rome, July 27, 1938]

Your Excellency,

I turn to you, the supreme inspirer and moderator of Justice, so that the measures most suitable for the recovery of my son, Ettore Majorana, may be intensified to the extent possible. He was a professor of theoretical physics at the University of Naples, a post to which he had been appointed for exceptional merits last November.

His painful and sudden disappearance has by now lasted for four months and there has been only one sure trace of him. In the last days of March or early April, Ettore Majorana appeared in a very agitated state at the Superior of the Church of Gesù' Nuovo, in Naples, and asked to be hosted in a retreat to experiment with religious life.

Having not been immediately accepted, for obvious reasons, he disappeared and there has been no news of him since. All of the investigations carried out by the Ecclesiastical Authority have been fruitless.

He was always wise and sensible, and the drama in his soul or nerves seems to be, in any case, a mystery. But one thing is certain, as all his friends, family, and I myself as his mother will attest: one never saw in him any moral or medical evidence that would suggest suicide; on the contrary, the serenity and seriousness of his life and his studies allow, indeed require, that he be considered solely a victim of science.

And of this there is no better witness than the Royal Academy of Italy member S.E. Fermi, who was his mentor and friend and who wanted to address to Your Excellency the attached letter, as an expression of the esteem he has for my son.

It is my understanding that the police have diligently endeavored in the search, unfortunately thus far without success.

If I am permitted an opinion, the search for my son should start in the countryside, in some farmer's house where it is far easier to escape the vigilance and careful searching of the police, and where he can conserve for quite some time the few thousand lire that he brought with him.

But so far there have been no reports, even though the search bulletin indicated him three times. In the event that my son is abroad, I make known to Your Excellency that his passport (No. 194925) expires in August and must be renewed at a consulate.

Your Excellency, it is an illness caused by his worthy studies, perhaps completely curable but possibly destined to worsen beyond remedy if neglected; your powerful intervention may determine the fate of this search and the life of a man.

Your Excellency, to whom we attribute the most ingenious and generous initiatives inspired by enlightened understanding and crowned by triumphant success, before you kneels a distraught mother, but one full of faithful hope.

(Dorina C. Majorana).

T/8 — Letter from Enrico Fermi to Mussolini (27th July 1938)

Rome, 27 July 1938 XVI

Duce,

with reference to the letter of the Majorana family which I have attached to this present letter, allow me to describe what is in my opinion the scientific position of Ettore Majorana and what are the future prospects of his work if, as all his colleagues so anxiously desire, he may be returned to his work for Italian Science.

I have no hesitation in telling you, and I do not exaggerate, that among all the Italian and foreign scholars I have ever known, Majorana, for his depth of intellect, is the one that most impressed me.

Capable at the same time of performing daring hypotheses and criticizing his own work and that of others; highly expert and profound mathematician who, beneath the veil of figures and algorithms, never lost sight of the real essence of the physical problem, Ettore Majorana possessed, to the highest degree, that rare combination of skills that make up a theoretical physicist of 'gran classe'. And indeed, in the few years that he had available to carry out his work, he gained the attention of scholars around the world, who recognized him as one of the most powerful intellects of our time and a guarantee that more breakthroughs would follow.

For these reasons, when a few months ago the work of Majorana earned him the highest recognition in his appointment as professor for exceptional merit, the decision was universally applauded by all interested parties. And the subsequent news of his disappearance has dismayed those who see in him someone who still has much to contribute to the prestige of Italian Science.

I am certain that I represent the unanimous sentiment of all scholars in expressing the desire that the search may soon lead to the discovery of Majorana and return him to the affection of his family and to his great work.

Please accept my expression of deep devotion,

Enrico Fermi.

10

Other Documents Related to His Professorship and His Disappearance

D/ME1 — Minutes of the Board of Examiners for the Professorship (25th October 1937)[1]

Selection Committee to the Chair of Theoretical Physics at the Royal University of Palermo

Minutes No. 1
The Commission appointed by S.E. the Minister of National Education, and consisting of Professors Antonio Carrelli, S.E. Enrico Fermi, Orazio Lazzarino, Enrico Persico, and Giovanni Polvani assembled at 4 o'clock p.m. on October 25th, 1937 in a lecture hall of the Institute of Physics at the University of Rome.

The Committee has constituted itself with S.E. Fermi appointed as President, and Carelli as Secretary.

After a thorough exchange of ideas, the Commettee was unanimous in recognizing the highly exceptional scientific

---

[1] This and following documents are from the State National Archive: Series "Directorate General for Higher Education"; folder "Personal – Series II"; file "Ettore Majorana".

standing of Professor Ettore Majorana, who is one of the candidates. Therefore, the Commission decided to send a letter and a report to S.E. the Minister to propose the opportunity to appoint Majorana a professor of theoretical physics for high and deserved reputation in a University of the Kingdom, independent of the competition required by the University of Palermo.

The Committee, awaiting instructions from the Minister, is adjourned until a new convocation.

The assembly is concluded at 7 o'clock p.m.

Read, approved and signed here,

Enrico Fermi
Orazio Lazzarino
Enrico Persico
Giovanni Polvani
Antonio Carrelli.

D/ME2 — Letter from the Committee to Minister Giuseppe Bottai (25th October 1937)

Rome, October 25, 1937 XV

To His Excellency The Minister of National Education
ROME

Your Excellency,

The Selection Committee to the Chair of Theoretical Physics of the Royal University of Palermo, now constituted, in conducting its operation was unanimous in identifying, after a thorough exchange of opinions, that among the nominees, Prof. Ettore Majorana has a national and international scientific position of such resonance that the Commettee hesitates to apply to him the normal procedure for university competitions

We therefore propose to Your Excellency the opportunity to appoint Majorana, for high and deserved reputation, Professor of Theoretical Physics in one University of the Kingdom, regardless of the competition asked by the Royal University of Palermo.

We are honored to attach a report on the scientific activity of Majorana.

Respectfully yours,

Enrico Fermi
Orazio Lazzarino
Enrico Persico
Giovanni Polvani
Antonio Carrelli.

D/ME3 — Report on the Scientific Activity of Prof. Ettore Majorana

Report on the scientific activity of Prof. Ettore Majorana:

Prof. Ettore Majorana graduated in physics in Rome in 1929. Since the beginning of his scientific career, he has demonstrated a depth of thought and conceptual genius that has attracted the attention of scholars of theoretical physics around the world.

Without listing his works, which are all quite remarkable for the originality of the methods employed and the importance of the results achieved, we limit ourselves here to the following notices:

In modern nuclear theory the contributions made by this researcher with the introduction of forces known as "Majorana Forces" are universally recognized as being, among the most important, those which allow us to understand theoretically the reasons for the stability of nuclei. Majorana's work is today the basis for the most important research in this field.

In atomic theory, Majorana deserves the credit for having solved, with simple and elegant considerations of the symmetries, some of the most intricate questions regarding the structure of the spectra.

In recent work he devised a brilliant method for treating positive and negative electrons symmetrically, finally eliminating the need for the highly artificial and unsatisfactory hypothesis of an infinitely large electric charge diffused throughout space, a question which had been unsuccessfully tackled by numerous other scholars.

Enrico Fermi
Orazio Lazzarino
Enrico Persico
Giovanni Polvani
Antonio Carrelli.

D/ME4 — Announcement of Appointment Out of Competition

MINISTRY OF NATIONAL EDUCATION
Directorate General of Higher Education
Division I Position 23 Rome

To the attention of Prof. Ettore Maiorana[2]
Viale Regina Margherita 37
ROME

SUBJECT: Announcement of appointment

We hereby communicate that, pursuant to art. 8 of Leg. Decree n. 1071, June 20, 1935 XIII, your appointment has been settled,

---

[2] *Sic.*

regardless of the normal competition procedure, as a professor of theoretical physics at the Faculty of sciences of the Royal University of Naples, for the high fame of singular expertise which you have attained in the field of studies regarding that discipline, with effect from 16 November 1937 XVI.

THE MINISTER.[3]

D/MEJ — Matriculation Status (20th January 1938)

MATRICULATION STATUS

Surname and name: Majorana Ettore

Son of: the late Fabio *and of*: Salvatrice Corso

Born in: Catania *on the day*: 5 August 1906

If never married, espoused or widower: never married

APPEARANCE:	HEALTH:
brown, thin, medium build	somewhat delicate

EDUCATION:

Degree in Physics

Foreign languages

Can speak or read:	*Can write:*
German, English, French	French and German (technical matters)

Ettore Majorana

January 20, 1938-XVI.

[3] For the answer to this notice, see the Letter *MM/N*, Chapter 9.

D/ME6 — Letter from Carrelli to the Rector (30th March 1938)

R. UNIVERSITY OF NAPLES

Naples, March 30, 1938-XVI

STRICTLY RESERVED AND CONFIDENTIAL

Dear Rector,

It is with great sorrow that I communicate to you the following:

On Saturday, March 26th at 11 in the morning I received an urgent telegram regarding my colleague and friend Ettore Majorana, professor of theoretical physics at this University, conceived in these terms: "Do not be alarmed. A letter will follow. Majorana". *This missive was incomprehensible to me. I inquired and found that on that morning he had not conducted his lesson. The telegram came from Palermo.*

With the 2 o'clock p.m. delivery of the mail, I received a letter from Naples dated the previous day, in which he expressed suicidal intentions. I understood then that the urgent telegram from Palermo the following day was in fact intended to reassure me, to provide me proof that nothing had happened. In fact, Sunday morning I received a letter by express mail from Palermo, in which he told me that those bad feelings had disappeared and that he would return soon.

Unfortunately, he did not appear the next day, Monday, neither at the Institute nor at the hotel where he had been staying. A bit alarmed at his absence, I passed the news of what had happened to those of his family who were residing in Rome. Yesterday morning, his brother [Salvatore] *visited me, and he and I went to the police Superintendent of Naples, requesting that he inquire with police headquarters in Palermo as to whether Prof. Majorana was still staying at a hotel in that city. Since, as of this morning, I have still not received any news, I inform you of what has occurred in the hope that my colleague has simply sought some respite following a period of exhaustion, of unhappiness, and will soon be back among us to offer his important contributions of work and intellect.*

With observance,
Antonio Carrelli.

D/PP — Letter from Giovanni Gentile to the Chief of Police, Bocchini (16th April 1938)

Urgent
To S.E.
The Sen. Arturo Bocchini
By His Hands

SENATE OF THE KINGDOM

Rome, 16.04.1938 XVI

Dear Excellency,

I beg you to please receive and listen to Dr. Salvatore Majorana, who needs to confer with you regarding the unfortunate case of his brother, the missing professor.

A new lead would seem to warrant a new investigation, in the convents of Naples and its surroundings, perhaps throughout all southern and central Italy. I highly recommend this to you. Prof. Majorana has been in recent years one of the leading Italian scientists. And if, as it is hoped, there is still time to save him and return him to his normal life and to science, we must leave no stone unturned.

Cordial greetings and best wishes for Easter,

Giov. Gentile[4]

D/ME7 — Letter from the University Rector of Naples to the Minister (29th April 1938)

R. UNIVERSITY OF NAPLES

To the Hon. Ministry
of National Education
Directorate General of Higher Education
ROME

Subject: Prof. Maiorana[5] *Ettore*

HIGHLY CONFIDENTIAL

Following my previous communications, I transcribe below the note of April 29 n. 87966 of the Police Superintendent of Naples:

"As Your Excellency knows, in March Prof. Ettore Maiorana[5] *left the Hotel Bologna in this city, indicating his suicidal intentions to Prof. Carrelli of the Institute of Experimental Physics at your distinguished University — intentions which he did not carry out.*

---

[4] From the Archivio Centrale dello Stato (Rome); Series "Polizia Politica: Personali"; envelope no. 780; file "Ettore Majorana". For the origin of the other documents reported in this Chapter 10, see Chapter 1.

[5] *Sic.*

At the request of his brother, Dr. Salvatore, a search began and was subsequently intensified, but it has been thus far unsuccessful. It only emerged that the missing person, apparently on the 12th, appeared at the Convent of St. Pasquale di Portici requesting admittance to that religious order, but the request was denied, so he left for an unknown destination.

The investigation continues with diligence and, if the results are favorable, we will inform Your Excellency"

THE RECTOR.

D/ME8 — New Letter from the Rector to the Ministry (28th July 1938)

Naples, July 28, 1938-XVI

R. UNIVERSITY OF NAPLES
protocol no.: 941
Position: *Reserved*

To the Hon. Ministry
of National Education
Directorate General of Higher Education
ROME

Subject: *Prof. Ettore Maiorana*[5]

The local police headquarters, with sheet no. 138008 of the 24th of the current month, with reference to my renewed urgency, informs me that the very active search involving all the police headquarters of the Kingdom conducted thus far to find Prof. Ettore Maiorana[5] *have failed, and that they shall report any useful updates to me.*

I confirm the assurance already given to you, that I will keep you informed of any possible news I become aware of with regard to the aforesaid.

THE RECTOR.

D/ME9 — Decree of the Minister Giuseppe Bottai (6th December 1938)

THE STATE SECRETARY MINISTER
OF NATIONAL EDUCATION

Having viewed the combined disposition of art. 109 of the Consolidated Act of legislation on higher education approved by Royal Decree no. 1592 31, August 1933, XI, and art. 46 of the Royal Decree no. 2960 of December 30, 1933, XII;

Considering that Prof. E. *Majorana,*5 *professor of theoretical physics at the Royal University of Naples, was absent from his office without justification for a period longer than ten days starting from the March 25 1938 XVI;*

Considering that, despite the search conducted, no information regarding the aforementioned professor has been obtained:

DECREES:

As of March 25th, 1938 XVI, Prof. Ettore MAIORANA,[5] *professor of theoretical physics at the Royal University of Naples, is dismissed from his post.*

This Decree shall be communicated to the Court of Auditors for recording.

Rome, December 6th, 1938-XVI

MINISTER
Bottai.

11

Other Testimonies

T/9 — Testimony of Giuseppe Cocconi (18th July 1965)

Geneva, 1965, July 18th

Dear Amaldi,[1]

In a discussion that took place long ago in the book [later published by the Accademia dei Lincei] *that you are writing about Ettore Majorana, I told you that I, too, had tenuous contact with Majorana just prior to the his disappearance.*

You expressed then the wish that I could describe my recollections in greater detail, so here I will try to satisfy you.

After having just graduated in January 1938, I was offered, mainly by you, the opportunity to come to Rome for six months as an assistant at the university's Institute of Physics. Once there, I was fortunate enough to join Fermi, Bernardini (who had got a teaching post in Camerino a few months prior) and Ageno (he a young graduate as well), to engage in research regarding the products of the disintegration of "mu mesons" (then called mesotrons or yukons) produced by cosmic rays. The existence of "mu mesons" had been proposed about a year earlier, and the problem of their decay was already quite fashionable.

---

[1] Testimony sent to Edoardo Amaldi, from *CERN* in Geneva, by the distinguished Italian physicist G. Cocconi.

It was indeed while I was with Fermi in the small workshop on the second floor, he working intently at the lathe on a piece of a Wilson chamber that was intended to reveal the mesons in their end range, and I busy building a jalopy to illuminate the chamber using the flash produced by the explosion of a strip of aluminum on a shorted battery, that Ettore Majorana came in looking for Fermi. I was introduced to him and we exchanged a few words. A dark face, and that was all.

It would have been a quite forgettable episode if, a few weeks later while with Fermi in the same workshop, I had not heard the news of Majorana's disappearance from Naples. I remember that Fermi busied himself by phoning various places until, after a few days, one got the impression that no one would ever find him.

It was then that Fermi, trying to impress upon me the significance of this loss, and who was so severe when judging others, expressed himself in a rather unusual way. And at this point, I would like to repeat his words just as they echo in my memory:

"Because, you see, in the world there are various categories of scientists, people of second and third rank, who do their best but do not go very far. There are also people of the first rank, who make discoveries of great importance that are fundamental for the development of science" *(and here I have the distinct impression that he would have put himself in this category)*. But then there are the geniuses like Galileo and Newton. Well, Ettore was one of these. Majorana had what no one else in the world has. But unfortunately, he lacked what is instead common in other men, plain good sense."[2]

I hope these lines provide you with what you wished to know.
Sincerely,

Giuseppe Cocconi.

T/10 — Letter from Bruno Pontecorvo (9th April 1975)

Dubna, 09.04.1975

Dear Professor Recami:
Thank you very much for your letter of March 26, 1975. I am very pleased to tell you that the material on Majorana you sent me earlier, and especially what you sent me along with your last letter, was of great interest to me, for which I thank you heartily.

I am to understand that the material in question is sent to me personally. If, in time, this material is either wholly or partially published in Italy, I would be grateful if you let me know, as I could write an informative article on Majorana for a few Russian newspapers, which certainly would be of great interest to Soviet readers.

Thank you also for your interesting article.

I wish you success in your scientific research as well as your investigations in the history of physics.

My cordial and repeated thanks,

Bruno Pontecorvo
11.IV.75.

---

[2] See Chapter 2.

T/11 — Letter Regarding Bruno Touschek (29th October 1976)

Rieti, 29.10.76

Dear Recami

... We discussed with Touschek your work on Majorana in Scientia *110 (1975) 577; he made some remarks about your list of Majorana's scientific works on page 585. In his view, you should add the theory of "The Majorana oscillator" which is implicitly contained in his theory of the neutrino.*[3] *The Majorana oscillator is described by an equation similar to* $q + \omega^2 q = \in \cdot \delta(t)$, *where* $\in$ *is a constant and* δ *is the Dirac delta function.*[3] *According to Touschek, the properties of this oscillator are of considerable interest, especially as regards its energy spectrum. There is not, however, a bibliography for this and I hope to provide more details as soon as Touschek explains it better to me.*

Yours sincerely,

Eliano Pessa.

T/12 — Testimony of GianCarlo Wick (16th October 1978)

Pisa, 1978, October 16th

Dear Prof. Recami

Please excuse me if I haven't answered your letter dated Sept. 16 until now. But I arrived in Pisa recently, and I also had to give

---

[3] See Chapter 5.

my complete attention to boring practical issues related to housing here in Pisa.

The scientific contact [between myself and Ettore] *which Segré mentioned to you did not take place in Leipzig, but in Rome at the Volta Congress (thus well before Majorana's stay in Leipzig). The conversation took place at a restaurant, in the presence of Heitler, and therefore without chalkboard or written formulas. But despite the absence of details, what Majorana described orally was a "theory of relativistic charged particles with zero spin based on the idea of quantization of fields" (second quantization). When much later I saw the work of Pauli and Weisskopf,*[4] *I was absolutely convinced that what Majorana had described was the same thing. Of course, Majorana published nothing and probably did not speak to many people. I have absolutely no reason to think that Pauli and Weisskopf knew anything about it. I never had the opportunity later to speak to Heitler of this episode, to see if he remembered. One should not be surprised if he had forgotten it, because Majorana had spoken of the matter with that detached and ironic tone that he often used even regarding his own matters. In short, without giving them importance.*

I do not know if these somewhat succinct memories will be of any use, but these things occurred a very long time ago and if I try to add details I risk imagining things. Memory can play funny tricks…

Cordially yours,

G.C. Wick.

---

[4] See also Testimony *T/4* of V. Weisskopf, Chapter 9.

T/13 — Letter from Franco Rasetti (4th January 1979)

Waremme - Belgium
04 January 1979

Dear Professor Recami

Your letter of 07.12.78 sent to me at the Johns Hopkins University was forwarded to my new address and I only received it today.

Thank you for the interesting documents on Ettore Majorana that you sent me. The reduced photocopy of your writing is perfectly legible. It was complete news to me that you discovered a collection of letters by Majorana... For almost twenty years I have no longer been involved in physics, present or past, my studies being exclusively devoted to geology, paleontology and botany.

Unfortunately I know very little of Majorana, since he lived away from the other inhabitants of the Physics Institute of Rome, and I probably had less contact with him than anyone else; if he spoke to someone, it was rather with Amaldi or Segré, more or less his peers, or with Fermi, with whom he could discuss theoretical physics. Even if he tried to explain his theories, I would certainly have not understood.

So I can say that almost everything I know about Majorana's life I learned reading Amaldi's biography. Among other things, in the year of his mysterious disappearance, I was in the United States.

My health is always the same since April 1976 and I am too old for it to substantially improve.

Cordial greetings,

Franco Rasetti.

T/14 — Testimonies of Franco Rasetti (22nd June 1984)

Waremme, 22.06.1984

Dear Professor Recami,

I beg your pardon for replying only now to your letter of April 4th of last year. It arrived while I was away on a trip for botanical purposes from May 6th to June 17th, so I have had only a few days to look it over. At the moment I do not have the time to make all of the comments the subject deserves, so I'll relate to you only my first impressions while resolving to write you a more detailed account later.

I could not attend the "Fermian"[5] conference because I was ill for a month and a half, which caused me to abandon a trip to Palermo for which I had already purchased a plane ticket; a trip for the purpose of photographing the Orchidaceae characteristic of the province of Palermo. I made another trip, the one stated above, to the south-central part of Italy. As you know, many years ago I lost all interest in physics, devoting myself with great success and satisfaction first to Cambrian Paleontology of Canada, the United States and Sardinia, then to the Botany of the Alps, and finally to Italian Orchidaceae.

Even from a first, brief overview of your letter, I understand that you have information and knowledge of Ettore's life which is far more comprehensive and thorough than what has been written by others, who possessed much more imagination than seriousness.

---

[5] See Footnote 2 of Chapter 4 (and Footnote 6 of Chapter 2).

I'd like to point out an incredible "gaffe" made by the Corriere della Sera [of 13 Dec 83, page. 16] *in the caption under the famous photograph called the "three priests" (because a friend of Laura Fermi, seeing a copy hanging on the wall in her sitting room, exclaimed: "Who are those three priests?").*[6] *From the caption of the* Corriere, *it would seem that the photograph represents Fermi, Rasetti and Segré as undergraduates! Rather, Fermi and I graduated in 1922, when the Fascist regime did not yet exist and undergraduates did not wear black shirts! The photograph was instead taken in the mid-thirties, and all three of us were members of the graduation committee, not graduates ourselves! If such monstrosities are present in what should be considered a serious newspaper, what should we believe of what is written in less reputable journals?*

With this I conclude my comments, for the moment.

Kind and cordial regards,

Franco Rasetti

P.S. I neglected to mention that it would be vain to expect from me any information regarding Majorana that is not already well known. Of the physicists of Via Panisperna, I am certainly the one who knew him least. In any case, I was somewhat older than he, while Segré and Amaldi were more or less the same age. Furthermore, I was connected to them through sporting activities, while Majorana was totally foreign to any sport. I used to play tennis

[6] *Corriere della Sera* of 13th December 1986, page 16. The photograph in question is reproduced in this book.

with Segré and Amaldi, but mostly we were mountain climbing companions (for example, we completed several difficult climbs, always without a guide, including the crossing of the Matterhorn).

Waremme, May 1988

Dear Prof. Recami,

Please forgive me if I did not write earlier to thank you for sending me your fine book, Il caso Majorana. [The Majorana case]. *I read it with much pleasure and interest [...]*

Of course, I did not expect any more news surrounding the "disappearance", after such a long time from those events. But I was interested in the description of the life of Majorana in Leipzig and his relations with the local physicists [...]; where it seems that M. had become much more sociable than in Rome and took an active part in the Institute of Theoretical Physics, contrary to what happened in Rome.

Unfortunately, my physical and mental conditions deteriorated a lot after a brain attack that struck me two years ago in Rome, and brought me almost to death; [...] Nor there is any hope that it will improve in the future. I can go for small walks around Waremme, unfortunately not very interesting (there are mostly extensive wheat fields), but I cannot embark on any trips [...] Since then I can no longer drive a car, and have been getting along as best I can, hoping for an end soon enough to a life that is not worth prolonging much further.

I have lost a lot of memory, and this fact has also some advantages, namely, that I read with interest and pleasure even books that I have already read recently, no longer remembering the details; and among these, particularly yours.

Please accept, in addition to my gratitude, my most cordial greetings and best wishes for your scientific work,

Franco Rasetti.

T/15 — Testimony of Rudolf Peierls (2nd July 1984)[7]

Oxford, 2.VII.1984

Herrn Donatello Dubini
Köln, West Germany

Sehr geehrter Herr Dubini,

Ihr Brief vom 4.VI., der nach Cambridge adressiert war, hat mich jetzt erreicht.

Ich war mit Ettore Majorana in Fermis Institut in Rom zusammen, im Winter 1932-3. Er erschien mir als ein ausserordentlich begabter Physiker, etwas schüchtern, und dem Faschismus sehr entgegengesetzt. Das war bevor er seine berühmten Arbeiten über Kernkräfte und über Neutrinos schrieb. Ich habe diese Arbeiten natürlich mit grossem Interesse verfolgt, aber ich habe ihn nicht mehr gesehen, soweit ich mich erinnere.

Über sein Verschwinden habe ich wohl schon in 1938 gehört, aber ich weiss nicht mehr wann oder von wem. Natürlich hat

---

[7] For the translation, see Footnote 8 of Chapter IV.

das allen Physikern sehr leidgetan, aber wir wussten zu wenige Einzelheiten, um über die Ursachen zu spekulieren…

Mit besten Grüssen

Ihr

Rudolf Peierls.

T/16 — Letter from Emilio Segre (2nd October 1987)

Prof. Emilio Segre
Lafayette, California

02 October 1987

Dear Prof. Recami

I received you letters dated Sept. 12 from Campinas addressed to the Dep. of Physics and to the Radiation Lab, and I hasten to reply with a copy to Catania, in the hope that one of the two reaches you.

When your book Il Caso Majorana *was released in Italy, I got a call from* Il Giornale *inviting me to write about it. I had not seen the book and I replied that I could not write about a book I had not seen. From* Il Giornale *they dictated a brief summary of the contents of the book. I answered that I had already written an article for them regarding* OTHER *books on Majorana, which they published on Dec. 17, 1975, and that article was still valid. Some Italian friends have told me that* Il Giornale *has republished my 1975 article, but so far I have not seen the issue of the newspaper and so I do not know if, and how, they explained what it was in reference to. I saw an interview with you on* La Stampa *on the 29th August.*

In the meantime I received your interesting book from Mondadori. Much of the published material was unknown to me; in particular, the family correspondence and the letter and report of the members of the Committee for the Competition in Palermo in which they seek the appointment of Majorana for special merit, and all the material relating to Argentina. [...]

I disposed the publication, for historical interests only, of the letter addressed to me by Majorana from Leipzig, on May 22, 1933, with a few brief explanatory comments.

Very cordial greetings,

Emilio Segre.

T/17 — Calligraphy Examination (6th May 1972)

Milan, 06.05.1972

Dear Erasmo,

never before this moment has graphology revealed itself to be so useless, even disrespectful. I think that the recommendations of Cesare Pavese can easily be adapted to Majorana: do not engage in gossip. But when has there ever been respect for the solitary? The invective of Nietzsche ("Woe unto him who kills a solitary") has always fallen on deaf ears. I say this because last night's conversation has reinforced my previously-formed conviction that first and foremost Ettore Majorana was (and I don't exclude that he still is) of very high moral character and that he must have suffered greatly. Let us not forget the climate in which he obtained his amazing results and, in particular, how much it was fatuously euphoric the Italy of 1938. The "National Institute of

Statistics" (ISTAT) uses this year as the basis for the calculation of its indices!

Thank you for the material you have given me, definitely first-hand, but I must tell you that, rather than work on the lecture notes originals, I preferred to work — so to speak — on the photocopy of the letter dated February 23, and that's because I prefer a sheet without the square rulings on the paper. What did I find? A sense of futility, I repeat (because discovering Majorana's intelligence is like reinventing the wheel), and also a sense of irreverence ... Let's analyze the handwriting, in any case.

My first impression is that he is restless, never satisfied with himself or with others. *What worsens the picture is the second finding, which is that* his soul does not seem anchored in a divine entity or belief: *Quite the contrary, I note in the use of certain letters (the "g" for instance) a* reluctance to see over and above things. *I reconfirm that the subject is fundamentally* a positivist, and gifted with a *stringent, consequential logic. Third surprise:* introversion *(small volume of the words), which is fed by a vast background of* skepticism *(angled inclinations) and* pessimism *(the falling space between one line and another). In short, we have up to this point a picture of a person who is somewhat "gattopardian", influenced by a well-defined class and secular position. Add to this that he is a restless soul tormented by fears and doubts which he does not* want *(I emphasize the word because the subject has* strong will*) to evade. Do you understand? Natural logic combined with skepticism force him into genuine doubts and torments which he will not wish to shirk at all, if he* *cannot overcome them with logic. A tragic closed circle that only a purpose, a goal, an authentic teleology could dissolve.*

Allow me to make a few comments on the overall appearance of the letters. Very good style, elegant, cultured, which fits his physical appearance (his phrases recoil from the rhetorical effects in the same way that he pathetically shrinks from the lens that depicts him)... They also reveal a love for precision, a strong, deep and ineradicable desire for clarity and logic. And is the quote about the "Ibsenian girl" just a cultural habit, or deep down does it reveal that he fears the discovery of his desire for renewal? We talked about the sea that "refused" him and of the expression "I will hold dear memories, at least until eleven o'clock"; a see-saw of emotions (the tango songs of that age are full of the boy and girl who come, go, return, run away,...) or rather a sense of panic over a deadline, of a duty from which he cannot escape? As you see, we cannot pore over someone without starting a discourse that will be unavoidably too long... But I can say this: Majorana must have been a gentle and good person, in need of affection more than ever, *and I think the best eulogy one could give him is to approach his vicissitude with respect and understanding.*

Gianni Sansoni.

12

The Possible Presence in Argentina: Testimonies

T/A1 — Testimony of Tullio Regge (28th November 1978)

ACCADEMIA NAZIONALE DEI LINCEI

Turin, 28 Nov. 1978

Dear Recami

I answer you late because you wrote to Princeton, rather than to Turin where I have returned.

The colleague from Tel Aviv "are" two. One is Yuval Neeman, who gave me an inexact version but one that gained my attention.[1] *Y.N. had heard it at a party in Texas hosted by Wheeler, W. learned of it in Varenna*[2] *(I believe) from Robert — now Yehuda — Meinhardt, who is now Israeli, but was previously Chilean. R. Ruffini knows him also. R.Y.M. told me personally, while I was in Tel Aviv, that Carlos Rivera, a professor at the Catholic university in Chile, had an experience in Argentina similar to that which later appeared in* "Oggi." *By chance, shortly afterward I went to Chile and I made a point of speaking with Rivera, who is truly the key person in the whole*

[1] See Chapter 8.

[2] Varenna on Como Lake together with Erice represents one of the most important Italian centers for International physics conferences.

affair. Others have reported stories more or less inexact with respect to Rivera's accounts, which were subsequently published in "Oggi". Published, I would say, without alteration. Rivera confirmed the meeting in Buenos Aires with Tullio Magliotti's mother (Mrs. Talbert), during which she'd stated that her son was a friend of a certain physicist called Majorana, who wrote formulas and quite disliked Fermi (reciting from memory) and who left Italy for this reason.[1] *What is strange is that Rivera also said he had met a chef at the Hotel (Continental?) who more or less told him a similar story. What can I say?*

Rivera certainly did not seem like a pathological liar. He is a respected professor at Catholic University, educated at Gottingen, of high culture and does not seem the type to tell lies. I'd recommend that Miss Majorana contact C. Rivera directly at Catolica *in Santiago, Departamento de Física… I don't know what else. I was struck by Rivera's obvious hostility toward Fermi, and he attributes hostility toward F. to the presumed Majorana. Rivera also has a theory.*

Rivera's theory:

1) *E. Majorana really was in Buenos Aires at that time.*
2) *It is a fact that Magliotti was anti-Perón and that many people disappeared, having been kidnapped and murdered by Perón's police. In his opinion, Magliotti and his mother were victims of this. There is no longer any trace, according to Rivera, of the two. T.M. was an engineer.*
3) *E. Majorana would have also been involved in T.M.'s political affairs and would have (?) suffered the same fate.*

I've reported these things just for completeness, and do not assume any responsibility for them. It would be better for you to speak with Rivera and/or hire someone to quietly conduct a series of investigations in Buenos Aires.

Best regards,

T. Regge.

T/A2 — Testimony of Yuval Neeman (20th October 1980)[3]

THE UNIVERSITY OF TEXAS AT AUSTIN
Centre for Particle Theory (CPT)
AUSTIN, USA

October 20, 1980

Dear Erasmo:

Your Majorana paper and the answers to my queries to you have all caught up with me.

My interest in Majorana was first awakened by conversations with the late Racah, who told me about the tragedy. In 1975 I learned about the renewed interest in Italy after Shasha's novel (I know this is not the correct spelling but I don't have it before me, so I write the name phonetically[4]*). I also read Amaldi's articles answering Shasha, etc.*

I am responsible for the revival of the Argentine version. Wheeler first heard it from Meinhardt (a Chilean Jewish Physicist who had meanwhile migrated to Israel) at Varenna in 1977, but

[3] Per la traduzione si veda il Capitolo 8.

[4] Sciascia.

confused Carlos Rivera[5] *with Saavedra. Realizing the importance and freshness of the issue in Italy, I related it to Regge, who had somebody call up Saavedra in Santiago. Saavedra said it wasn't him. I looked up Meinhardt in Israel and had him meet Tullio [Regge] when Tullio visited me there in May, 1978. Tullio went to Chile from Tel Aviv and got the details from Carlos Rivera. When I wrote to you, I did it after Tullio told me had reported it all to you, and I wanted to have the precise results of his inquires.*

I would certainly like to help check on these facts, but it seems very hard. The Argentine has had so many upheavals!

However, I shall try and find out whether the Jewish family, mentioned by Rivera as having been in contact with Majorana at the time, hasn't ended up in Israel. If I do discover something, I shall certainly let you know (and Miss Majorana) immediately.

With kind regards,

Yuval
(Yuval Neeman).

T/A3 — Testimony of Carlos Rivera (18th October 1978)[6]

PONTIFICIA UNIVERSIDAD CATÓLICA DE CHILE

Santiago, 18 October 1978

Dear Dr. Recami:

Today I received your kind letter in which you ask if what the journalist Mr. Gullace wrote in the magazine "Oggi", no. 41, of

---

[5] In the text, erroneously, Ribeira.

[6] For the translation, see Chapter VIII, Footnote 3.

14th October 1978, pp. 95–97, corresponds to what I told him by phone.

I can tell you that what Mr. Gino Gullace wrote does indeed correspond to what I know of the fate of Ettore Majorana.

I have no further information beyond that given to Mr. Gino Gullace.

I can assure you that Mrs. Talbert lived in terror in her apartment in Buenos Aires due to the oppressive regime of Perón.

If any other information comes into my possession in the future, you will be notified immediately.

With many kind regards,

Carlos Rivera C.
CARLOS RIVERA CRUCHAGA
DIRECTOR
INSTITUTO DE FÍSICA.

A4 — Letter from Carlos Rivera to Maria Majorana[7] *(28th November 1978)*

PONTIFICIA UNIVERSIDAD CATÓLICA DE CHILE

Santiago, 28 November 1978

Most esteemed and respected Ms. María Majorana[8]*:*

I received your letter. I ask sincerely that you forgive me for the delay, I have many activities which I must see too.

---

[7] For the translation, see Footnote of Chapter 8.

[8] For phonetic reasons, here and in the following text, Rivera's secretary erroneously types Mayorana.

I absolutely do not want to be put back in contact with the journalist Gullace, nor will I give any public statement.

Sadly, with so much time having passed, I do not remember further details about Mrs. Talbert, neither her appearance nor the face of the waiter at the Hotel Continental in Buenos Aires.

In my opinion, the terror that Mrs. Talbert displayed was due to the horrible persecution that she had been subject to under Perón, and I am almost certain that her son died a victim of this persecution. It is not improbable that Dr. Ettore Majorana[8] *escaped the Perón persecution, and personally I do not know his subsequent fate (did he return to Italy?).*

Mr. Erasmo Recami wrote to me from Catania, but I cannot remember the name or the appearance of the waiter.

Please forgive me if I cannot ease your anxiety, since so many years passed that I am unable to provide you more details. But a doubt has come to me; that some wretched person could have passed himself off as the great physicist Ettore Majorana, pretending to be your brother in order to profit from his immense prestige. This suspicion is not entirely unfounded as the very name Majorana is often used by false Physicists.[9]

These things happen occasionally in history when the true circumstances of a disappearance are unknown.

The hypothesis that the name was false is probably the most plausible.[9] *Mrs. Talbert was of advanced age (she was a friend of my mother, who had met her on a trip from France to Argentina); she was quite sure it was Ettore Majorana, of whom she was*

[9] See, instead, Chapter 8.

quite fond. Despite the fact that she had never met him when he worked with Fermi in Italy, she had so much confidence that everything seemed certain to her.

For any other information that I receive which might be useful, you can count on me, since I will not pass any information to the press. And if I had known that Ettore had a sister, I would have written directly to you, without rendering any of it public domain.

With much affection and warmest greetings,

Carlos Rivera C.
CARLOS RIVERA CRUCHAGA
DIRECTOR
INSTITUTO DE FÍSICA
SANTIAGO, CHILE.

T/A5 — Letter from Giulio Gratton (5th June 1979)

UNIVERSIDAD DE BUENOS AIRES

Buenos Aires, June 5, 1979

Dear Professor Recami,

I am writing in response to your letter dated December 5 1978. I'm sorry to have to inform you that the investigations that I have carried in my faculty have not produced any results. In particular, no one recalls a certain engineer Tullio Magliotti whom you mentioned... There is currently no Hotel Continental,[10] *but I cannot exclude that there was a hotel with that name in the past.*

[10] Hotel Continental does indeed *exist* and it is one of the best known hotels in Buenos Aires!

In the Buenos Aires' telephone directory there are many different listings of Maiorana (with an "I", but this is not strange as some foreign names are changed in order to maintain the original pronunciation), but there are none with the same initial.

With this, I believe that I have satisfied your request, as much as I could, within the limits of confidentiality that you requested. To go further I think you would have to employ the services of a research agency, possibly through the consular authorities, and receive more detailed information about any acquaintances and friends to contact.

I take this opportunity to extend my most cordial greetings,

Giulio G.
Prof. Julio Gratton
Director
Departamento de Física.

T/A6 — Letter from Yuval Neeman (23rd November 1980)[11]

TEL AVIV UNIVERSITY
INSTITUTE OF ADVANCED STUDIES
PROFESSOR YUVAL NEEMAN, DIRECTOR

23 November 1980
YN/722

Dear Erasmo

To enable me to pursue the matter of Majorana's "Argentine" version, could you send me a copy of the relevant documentation

[11] Cf. also Chapter 8.

you mention in your letter of the 19th September 1980, including Tullio Regge's report on his conversation with Carlo Rivera?

Best regards,

Yuval.

T/A7 — Testimony of Blanca Asturias (7th April 1985)[12]

BLANCA DE MORA Y ARAUJO DE ASTURIAS

Paris, April 7 1985

Mrs. Maristella Fracastoro – Inst. De Physique – UNICAMP:

Dear Madam,

Several years ago I received a letter regarding the case of Professor Ettore Mayorana [sic] *and his friendship with my friends Cometta-Manzoni.*

I replied that my friend Eleonora C. Manzoni was certainly a friend of Mayorana,[13] *but that she had already been deceased for several years. Eleonora and her sister lived at the time in via Santa Fé 2189 Buenos Aires, Rep. Argentina. Maybe that there they can give you their new address. The other sister, married to a Venezuelan engineer, lives in Caracas, Venezuela: Lolò* [Lilò] *Cometta-Manzoni de Herrera is a professor of Literature at the University of Caracas, but I don't have her address because I knew where her home was, without needing an address. But it*

[12] Cf. also Footnote 11 of Chapter 8.

[13] The "Maiorana" (or Mayorana) spelling is due to phonetic reasons.

will not be difficult to find her because they are very important people there. Or, as a last resort, you can speak to my sister Lila de Mora y Araujo de Gándara Casares, Avenida …, Buenos Aires; her friendship with these sisters was as close as mine, but I do not know their new address.

I am sorry, Mrs. Fracastoro, that I am not able to be of more use. I, too, am far away from my country, where indeed I hope to return after the succession of the estate of my husband Miguel Angel Asturias, Nobel Prize for Literature, Grand Officer of the Legion of Honor of France, Grande Croce di Bolivar in Colombia and Grand Price of the Peace.

When wishing her success in her search, tell Mrs. Mayorana that her brother's name is not unknown to me, but we left Buenos Aires in 1961: Eleonora Cometta, a close lifelong friend who was dear to my heart, was still alive.

I advise you to write to my sister Lila de Gándara Casares.

Renewing my friendship with you,

Blanca de Mora y Araujo de Asturias

P.S.: while closing this letter I was fortunate enough to find the address of the sister of Eleonora Cometta-Manzoni (the mathematician), Lolò [Lilò] *Cometta-Manzoni de Herrera (Professor of Literature): calle …, Caracas, Venezuela.*

T/A8 — Letter from Leonardo Sciascia (10th September 1986)

Palermo, 10.9.86

Dear Professor Recami,

I read today your letter dated 23.5; and it is the first I am pulling from the pile that formed after four months of vacation (so to speak, I did actually do some work). I am only here for a brief stay, however: I am going to Rome and then, on the 20th, to Milan. In Rome, I'll be in the hotel Nazionale, Piazza Montecitorio; in Milan, at the Manzoni. I would like to meet you, if you happen to be there: also to talk about this work of yours, which I will read soon (a version compatible with the maculopathy that afflicts me).

I would like to make available to you certain letters that I received after the publication of my book (one that regards Argentina); the few that seemed to me rather sensible and reliable. I hope I don't have too much trouble finding them, given the disorder in which I am more and more immersed. I'd also like you to read the introduction to the German edition of my book — very, very interesting.

I'll be back in Palermo (but only to depart again soon thereafter: Palermo oppresses me), around October 15.

With the most cordial greetings,

Leonardo Sciascia.

List of Ettore Majorana's Publications

1. "Sullo sdoppiamento dei termini Röntgen ottici a causa dell'elettrone rotante e sulla intensità delle righe del Cesio", in collaboration with Giovanni Gentile jr., *Rendiconti Accademia Lincei*, vol. 8, 1928, pp. 229–233.
2. "Sulla formazione dello ione molecolare di He", *Nuovo Cimento*, vol. 8, 1931, pp. 22–28.
3. "I presunti termini anomali dell'Elio", *Nuovo Cimento*, vol. 8, 1931, pp. 78–83.
4. "Reazione pseudopolare fra atomi di Idrogeno", *Rendiconti Accademia Lincei*, vol. 13, 1931, pp. 58–61.
5. "Teoria dei tripletti P' incompleti", *Nuovo Cimento*, vol. 8, 1931, pp. 107–113.
6. "Atomi orientati in campo magnetico variabile", *Nuovo Cimento*, vol. 9, 1932, pp. 43–50.
7. "Teoria relativistica di particelle con momento intrinseco arbitrario", *Nuovo Cimento*, vol. 9, 1932, pp. 335–344.
8. "Über die Kerntheorie", *Zeitschrift für Physik*, vol. 82, 1933, pp. 137–145; "Sulla teoria dei nuclei", *La Ricerca Scientifica*, vol. 4 (1), 1933, pp. 559–565.

9. "Teoria simmetrica dell'elettrone e del positrone", *Nuovo Cimento*, vol. 14, 1937, pp. 171–184.
10. "Il valore delle leggi statistiche nella fisica e nelle scienze sociali" (published posthumously, G. Gentile jr., Ed.), *Scientia*, vol. 36, 1942, pp. 55–66.

Bibliography

E. Macorini (Ed.), *Scienziati e tecnologi contemporanei: Enciclopedia Biografica*, 3 vols., Mondadori, Milan, 1974.

E. Amaldi, *La vita e l'opera di E. Majorana*, Accademia dei Lincei, Rome, 1966.

E. Amaldi, "Ettore Majorana: Man and Scientist", in *Strong and Weak Interactions*, A. Zichichi (Ed.), New York, 1966.

E. Amaldi, "Ricordo di Ettore Majorana", *Giornale di Fisica*, vol. 9, 1968, p. 300.

E. Amaldi, "From the discovery of the neutron to the discovery of nuclear fission", *Physics Reports*, vol. 111, 1984, pp. 1–322.

E. Amaldi, in "Il nuovo Saggiatore", 4, Bologna, 1988, p. 13.

E. Arimondo *et al.*, "Ettore Majorana and the birth of autoionization", *Review of Modern Physics*, vol. 82, 2010, p. 1947.

M. Baldo, R. Mignani and E. Recami, "Catalog of unpublished manuscripts of E. Majorana", in *E. Majorana – Lezioni all'Università di Napoli*, Bibliopolis, Naples, 1987, p. 175.

M. Bunge, *La causalità*, Turin, 1970.

F. L. Cavazza and S. R. Granbard, *Il caso italiano: Italia anni '70*, Milan, 1974.

Conferenze e discorsi di Orso Mario Corbino, Rome, 1939.

D. De Masi (Ed.), *L'Emozione e la Regola: I Gruppi Creativi in Europa dal 1850 al 1950*, Laterza, Bari, 1989.

O. D'Agostino, "L'era atomica cominciò a Roma", *Candido*, vol. XIV (24–26), 1958. See also R. Mignani and F. Cardone, *Enrico Fermi e i secchi della Sora Cesarina*, Di Renzo, Rome, 2000.

F. and D. Dubini, *The disappearance of Ettore Majorana*, television program broadcast in 1987 (TV svizzera).

G. Enriques, *Via D'Azeglio 57*, Zanichelli, Bologna, 1971.

S. Esposito, "Covarinat Majorana formulation of electrodynamics", *Foundations of Physics*, vol. 28, 1998, pp. 231–244.

S. Esposito, E. Majorana jr., A. van der Merwe and E. Recami (Eds.), *Ettore Majorana — Notes on Theoretical Physics*, Kluwer, Dordrecht and Boston, 2003; Book of 512 pp.

S. Esposito, E. Recami, A. van der Merwe and R. Battiston, *E.Majorana — Unpublished Research Notes on Theoretical Physics*, Springer; Berlin, 2009; Book of 487 pages.

S. Esposito, "Majorana solution of the Thomas-Fermi equation". *American Journal of Physics*, vol. 70, 2002, pp. 852–856; "Majorana transformation for differential equations", *International Journal of Theoretical Physics*, vol. 41, 2002, pp. 2417–2426.

S. Esposito, *The Physics of Ettore Majorana*, Cambridge University Press, 2015.

G. Fraser, *Cern Courier*, 38, files n. 5 and 6 (Summer and September 1998).

M. Farinella, *L'Ora*, Palermo, 22 and 23 July 1975.

E. Fermi, "Un maestro: O.M. Corbino", *Nuova Antologia*, vol. 72, 1937, p. 313.

L. Fermi, *Atomi in famiglia*, Milan 1954.

C. Fontanelli, *Il caso Ettore Majorana: Aspetti storici e filosofici*, graduation thesis: mentor A. Pagnini (Fac. Lett. Filos., Univ. of Florence), 1999.

B. Gentile, "Lettere inedite di E. Majorana a G. Gentile jr.", in *Giornale critico della filosofia italiana*, Florence, 1988, p. 145.

E. Giannetto, "Su alcuni manoscritti inediti di E.Majorana", in *Atti IX Congresso Nazionale di Storia della Fisica*, F. Bevilacqua (Ed.), Milan, 1988, p. 173.

G. C. Graziosi, "Le lettere del mistero Majorana", in *Domenica del Corriere*, Milan, 1972.

F. e P.h. Gueret: French translation of the 1991 edition of the present book (unpublished).

G. Holton, *The scientific information: Case studies*, Cambridge, 1978.

C. Leonardi, F. Lillo, A. Vaglica e G. Vetri, "Quantum visibility, phase difference operators, and the Majorana sphere", preprint (Phys. Dept., Univ. of Palermo, Italy), 1998; "Majorana and Fano alternatives to the Hilbert space", in *Mysteries, Puzzles, and Paradoxes in Quantum Mechanics*, R. Bonifacio (Ed.), A.I.P., Woodbury, N.Y., pp. 312–315.

A. Majorana, "La questione degli spostati e la riforma dell'Istruzione Pubblica in Italia", speech to Parliament on March 11, 1899, Rome, 1899.

G., A., and D. Majorana, *Della vita e delle opere di Salvatore Majorana Calatabiano*, Catania, 1911.

E. Mignani, E. Recami and M. Baldo, "About a Dirac-like equation for the photon according to Ettore Majorana", *Lettere Nuovo Cimento*, vol. 11, 1974, pp. 568–572.

R. Penrose, *Ombre della mente* (Shadows of the Mind), Rizzoli, 1996, pp. 338–343 and pp. 371–375; "Newton quantum theory", in *300 Years of Gravity*, S. W. Hawking and W. Israel (Eds.), Cambridge University Press, 1987.

B. Pontecorvo, *Fermi e la fisica moderna*, Rome, 1972.

B. Pontecorvo, Contribution at the "Congress on the history of particle physics", Paris, 1982.

S. Ponz de Leon, "Speciale News: *Majorana*", TV broadcast on September 30, 1987 (*Canale Cinque*), for the interest of Arrigo Levi: htpps://www.youtu.be.com/watch?v=RLqHu2w7Rds [in Italian].

A. Ravelli *et al.*: website www.ilsegretodimajorana.it, with infos and some documents about R.Pelizza and the related "machine" [in Italian].

B. Preziosi (Ed.) *Ettore Majorana — Lezioni all'Università di Napoli*, Bibliopolis, Naples, 1987.

E. Recami: *In caso Majorana: Epistolario, Documenti, Testimonianze*, 1987 and 1991 [Mondadori and Oscar Mondadori, Milan, Italy]; and 2000, 2002, 2008, 2011 [Di Renzo Editore, Rome].

E. Recami, "The new documents on the disappearance of E. Majorana", *Scientia*, vol. 110, 1975, p. 577.

E. Recami, in *La Stampa*, Turin, June 1 and June 29 1975.

E. Recami, in *Corriere della Sera* (Milan), October 19 1982 and December 13 1983.

E. Recami, "E. Majorana: lo scienziato e l'uomo", in *E. Majorana — Lezioni all'Università di Napoli*, Naples, 1987, p. 131; "Ricordo di Ettore Majorana a sessant'anni dalla sua scomparsa: L'opera scientifica edita e inedita", in *Quaderni di Storia della Fisica* (S.I.F.), vol. 5, 1999; and in *Ciência & Sociedade*, *PERFIS*, Rio de Janeiro, 1997, pp. 107–172.

E. Recami: video of a talk [Palermo, 2015] presenting also some (brief) information about the known 1975–76 Controversy among L.Sciascia and Physicists like E. Amaldi and others (including ourselves): https://radioradicale.it/schada/459275/vi-leonardo-sciascia-colloquium-e-possibilmente-anche-dopo-19381975-2015-la-scomparsa?i=3486646 [in Italian].

E. Recami: video of a talk [Messina, 2018] presenting the originals of important documents on E. Majorana's life (and work), at an annual meeting of the Italian Soc. for Hist. of Physics and A.: htpps://youtu.be/toDKVYpI4TM.

V. Reforgiato, *Cenni biografici e critici su Angelo Majorana*, Catania, 1895.

V. Reforgiato, *Raccolta di recensioni e giudizi sulle opere del Prof. Avv. Giuseppe Majorana, Catania*, s.d.

A. Rocca, *Il Liberty a Catania*, Catania, 1984.

B. Russo, *Ettore Majorana — Un giorno di marzo*, television program first broadcasted on December 18 1990 (*Rai Tre – Sicilia*). In talian. See also the book with the same title (*Ettore Majorana – Un giorno di marzo*), Palermo, 1997: https://yiutu.be.com/watch?v=bv4DlzOTLKg , where l is the small letter el [in Italian].

G. Scavonetti, *La vita e l'opera di Angelo Majorana*, Florence, 1910.

E. Schrödinger, *Scienza e umanesimo*, Florence, 1970.

L. Sciascia, *I Catanesi com'erano*, Catania, 1975.

L. Sciascia, *La scomparsa di Majorana*, Turin, 1975.

E. Segré, *Enrico Fermi, fisico*, Bologna, 1971.

E. Segré, Una lettera inedita di E. Majorana, in *Storia contemporanea*, vol. 19, 1988, p. 107.

E. Segré, *Autobiografia di un Fisico*, Il Mulino, 1995.

B. Tarsitani, "O.M. Corbino", in *Sapere*, 49, Rome, 1983, n. 5.

C. Frank (Ed.) *The Farm Hall Transcripts — Epsilon Operation*, Institute of Physics Pub., Bristol, 1993.

S. Timpanaro, *Pagine di scienza: Leonardo*, Milan, 1926.

V. Tonini, *Il Taccuino incompiuto*, Rome, 1984 [interesting digression which, through a typical literary fiction, freely investigates the possible "secret life" of E. Majorana].

W. Wataghin, in *Boletím Informativo*, Instituto de Física Gleb Wataghin, Universidade Estadual de Campinas, Unicamp, Campinas, S.P., September 6 and 13, 1982.

F. Wilczek (NL), "Majorana returns", *Nature Physics*, vol. 5, 2009, p. 614.

J. Zimba and R. Penrose, *Studies in History and Philosophy of Science*, vol. 24, 1993, p. 697.

Appendix

*"In Memory of Ettore Majorana"**

1. *His Youth*

Ettore Majorana was born in Catania on 5th August 1906 to a distinguished family of professionals in that city. His father, Engineer Fabio Massimo (b. Catania in 1875, d. Rome in 1934) was the younger brother of Quirino Majorana (1871–1957), the renowned professor of experimental physics at the University of Bologna. Engineer Fabio Massimo had been Director of *Azienda Telefonica di Catania* for many years; he then moved to Rome where he was appointed Head of Division in 1928 and a few years later, Inspector General of the Ministry of Communications. The marriage of Engineer Fabio to Mrs. Dorina Corso (b. Catania, 1876, d. Rome, 1966), also from Catania, produced five children: Rosina, who later married Werner Schultze, Salvatore, a doctor of law and scholar of philosophy, Luciano, a civil engineer specializing in aeronautical construction but who

---

* Courtesy of Prof. E. Amaldi (first historian of Majorana) and the "Società Italiana di Fisica", we reproduce here his "In memory of Ettore Majorana" which appeared in the *Giornale di Fisica*, vol. 9, p. 300 (Bologna, 1968). Another version, enriched with scientific-technical considerations, had appeared in 1966 in Rome, published by the "Accademia Nazionale dei Lincei".

then devoted himself to the design and construction of instruments for optical astronomy, Ettore (born 5th August 1906 in Catania), and fifth and last, Maria, musician and piano teacher.

After completing his first elementary school years at home, Ettore entered as an intern at the *Istituto Massimo di Roma* [Massimiliano Massimo, recognized institute directed by the Jesuits], where he completed elementary school and continued on to middle school, which he passed in four years, having skipped the fifth. When in 1921 his family moved to Rome, he studied in the first and second year of Istituto Massimo grammar school as an external, but then for the third year shifted to the Torquato Tasso State School, where, in the summer term of 1923, attained his diploma with high marks [Italian: *w.* 7, *o.* 8; Latin: *w.* 7, *o.* 7; Greek: *w.* 7, *o.* 7; History and Geography: 8; philosophy: 7; mathematics: 9; Natural History: 7; Physics: 9; Gymnastics: 8].

In the autumn of the same year, Ettore joined the Biennium of Engineering Studies at the University of Rome and began to attend lectures and tutorials, regularly passing exams with very high scores.

Among his fellow students were his brother Luciano, with whom he spent much of his free time and shared mutual friends; there were also Emilio Segré, now a professor of physics at the University of California at Berkeley, and Enrico Volterra, now a professor of

Construction Science at the University of Houston in Texas.[1]

After the Biennium in Engineering, this group of young people, all very bright, began attending the School of Application for Engineers in Rome. Ettore continued to obtain high marks in all exams, except for one failure in hydraulics. As during the Biennium, so too at the School of Engineering Majorana acted as consultant to all his companions to solve the most difficult problems: especially if they were mathematical ones

During the period when he attended the School of Engineering, Majorana, like some of his classmates, began to tire of the way they were being taught some of the subjects; he believed that they dwelled too much on the detailed descriptions of the inessential, while not giving enough importance to the general synthesis, characteristic of a solid scientific framework. This firmly held belief was behind some of the frequent, lively and sometimes heated discussions he had with some teachers.

At the beginning of the second year of the School of Engineering (fourth from the beginning of their university studies), Emilio Segré decided to follow his old passion and shifted his studies to physics. This decision matured in him during the summer 1927, the period in which he had met Rasetti, then an assistant at the Institute of Physics of the University of Florence. Through

[1] Actually, it's *Austin* (capital city of Texas). (*Ed.*)

Rasetti, Segré also met the then 26-year-old Enrico Fermi who had recently been appointed (November 1926) adjunct professor to the Chair of Theoretical Physics at the University of Rome.

The creation of this new position was due to the work of O.M. Corbino, Professor of Experimental Physics and Director of the Institute of Physics of the University of Rome, who, having correctly assessed the exceptional skills of Enrico Fermi, undertook a series of steps to create a school of modern physics in Rome.

I also, having finished in June 1927 the end of the second year of the biennium for the study of engineering, decided to switch to physics on hearing an appeal by Corbino during a lesson, explicitly describing the stagnation of ideas that already existed in Europe in the field of physics and how the appointment of Fermi as professor in Rome in his view opened an exceptional window of opportunity for young people who had already begun to demonstrate that they were sufficiently equipped and were willing to undertake an extra effort in their study and theoretical and experimental work.

In the autumn of 1927 and early winter 1927–1928, Emilio Segré, in the new physics environment that had been formed in a few months around Fermi, spoke frequently of the exceptional quality of Ettore Majorana and at the same time tried to persuade Ettore Majorana to follow his example by pointing out how the study of physics was much more consistent than that of engineering with its scientific aspirations and its speculative ability. His move to physics took place at the beginning of 1928 after an interview with Fermi, the details of which can provide a good example of some aspects of the nature of Ettore Majorana.

He came to the Institute of Via Panisperna and was accompanied by Segré to Fermi's office, where Rasetti was also present. It was then that I saw Majorana for the first time. From a distance he looked slim, with a shy and uncertain gait. Up close you could see his jet black hair, dark complexion, slightly hollowed cheeks, sparkling and vivacious eyes: the overall appearance of a Saracen.

Fermi was then working on the statistical model [*of the Atom*] which later took the name of Thomas-Fermi. The discussion with Majorana immediately focused on current research at the Institute and Fermi quickly described the general ideas of the model, showed Majorana excerpts of his recent work on the subject and, in particular, the table which housed the numerical values of Fermi's so-called universal potential. Majorana listened with interest and, after asking for some clarifications, left without expressing his thoughts and intentions. The next day, in the late morning, he came to the Institute again, went directly to Fermi's study and asked, without preamble, to see the table that had been placed under his eyes for a few moments the day before.

Having received it in his hand, he drew a piece of paper from his pocket with a similar table he had calculated at home in the previous 24 hours. He compared the two tables, and finding that they were in full agreement with each other, said that Fermi's table was o.k. and then walked out of the studio and the Institute. After a few days he moved to Physics and began regularly attending the Institute.

2. *His University Studies in Physics*

Having crossed over to physics, Ettore Majorana quickly impressed everyone with liveliness of wit, depth of understanding and extension of studies that rendered him far superior to all his new fellow students. His critical spirit was also exceptionally penetrating and relentless, so much so that we nicknamed him the "Grand Inquisitor"; in the same context we jokingly called Fermi "The Pope", Rasetti the "Cardinal Vicar", and so on.

His ability to calculate was incredible. Not only did he perform complex numerical calculations completely by heart, but it only took him 20 or 30 seconds to perform alphanumeric calculations of rather complex definite integrals in his head; the same calculation would have taken a skilled mathematician a considerable number of steps: he was even able to substitute alphanumeric and numeric limits and directly give the final answer.

In 1928, during the months of May and June, which was the period for preparing and taking university examinations, we got into the habit of meeting before dinner, between seven and eight o'clock in the evening, at *Casina delle Rose of Villa Borghese.* In addition to Ettore Majorana, Giovanni Gentile jr., Emilio Segré, and I from the Institute of Physics, were Luciano Majorana, Giovanni Enriques, Giovanni Ferro-Luzzi, Gastone Piqué, all engineering students in the same year as Ettore. Sipping a drink or eating an ice cream, we would discuss our preparations for upcoming exams, or exams we had already taken, some of us physicists spoke about some results of atomic physics we had learned of recently, most of the time by Fermi, or one of the engineering students would talk about the properties of electromagnetic fields or some of its applications, or criticize their nemesis, the hydraulics professor. We also spoke about literature: Ettore generally knew and appreciated the classics and preferred the Shakespeare and Pirandello. We also spoke about various cultural issues, which Ettore was always vehement about, a little politics, but mostly about the Nobile expedition to the North Pole that had taken place right around that time (March–May 1928) and that had given rise to well-known complex human events.

The habit of going to the Casina delle Rose was continued, albeit with much less regularity, in the months of May and June of the following year, until we graduated.

Ettore Majorana, Gabriello Giannini (who later established himself as a builder and electronics manufacturer in the United States) and I graduated on the same day, 6th July 1929; Ettore presented a thesis on the mechanics of radioactive nuclei for which Fermi was his mentor, and scored 110/110 with honors. Reading this thesis even after almost 40 years continues to be striking for its clarity and the further discussions regarding the structure of nuclei and the theory of alpha decay.

After graduating, Ettore continued to attend the Institute, where he more or less spent a couple of hours in the morning and a few hours in the afternoon. This time was spent in the library where he studied mainly the work of Dirac, Heisenberg, Pauli, Weyl, and Wigner.

During that period his judgments on living scientists, including prominent ones, were almost always extremely severe, so as to give rise to the suspicion of a presumption and exceptional pride; his severity was less harsh or even non-existent when it came to his friends, but he implied equally severe judgments on himself and this was explicitly manifested in his work. The people close to him thus began to understand that such severity was nothing more than the manifestation of an unhappy and tormented spirit. Under an apparent isolation from others, not only physically but also emotionally, a lively sensibility hid which only rarely allowed him to forge friendships, but these were so profound as to reflect his region of origin.

On 12th November 1932 he was awarded a University teaching qualification (Libera Docenza) in theoretical physics: he only presented five pieces of work, but the committee made up of Enrico Fermi, Antonino Lo Surdo, and Enrico Persico was unanimous in recognizing *a complete mastery of theoretical physics* in the candidate.

3. *His Work in the Field of Atomic and Molecular Physics and the Evolution of His Interests in the Physics of Nuclei*

From the scientific output perspective, those years represent the first of two phases of the all too brief research activities of Ettore Majorana, which total nine works and a high popularization article. The first phase includes six works that all relate to problems in atomic and molecular physics; the second phase comprises only three questions concerning the physics of the nucleus or the properties of elementary particles.

The work belonging to the first phase can be further divided into three groups: The first consists of three pieces relating to atomic spectroscopy; the second group includes two that deal with some issues relating to chemical bonding. Finally, the third group consists of a single piece which deals with the issue of the non-adiabatic reversal of spin (spin-flip) in a beam of polarized atoms. All of these works have a striking impact for their very high quality: they reveal a deep understanding of the experimental data even in the most minute detail, and a rare ease, especially for that time, to exploit the symmetry properties of states to simplify problems or for the choice of the most suitable approximation to quantitatively solve individual problems; this latter ability without doubt derived, at least in part, from his exceptional calculative prowess.

In particular, works n. 2 and n. 4 [*see the list of Majorana's scientific notes after the end of this article*] allowed Majorana to become a leading authority on quantum theory for chemical bonding, a fact that would become crucial for future research. His thorough knowledge of the mechanism for the exchange of valence electrons, which is the basis of the quantum theory of homopolar bonding, later forms the basis for the hypothesis that nuclear forces are exchange forces.

Work n. 6 on spin reversal in a magnetic field is a classic issue associated with the handling of these matters, and as such is commonly cited: his results have subsequently formed the principle at the basis of the experimental method used to flip the spin of the neutron in a radio frequency field, a method used both in the analysis of polarized neutron beams, and in all the polarized neutron spectrometers used in the study of magnetic structures.

Majorana's interest in nuclear physics, which he had already manifested in his thesis, was strongly revived with the appearance of classic works that would lead to the discovery of the neutron at the beginning of 1932. In reality, his renewed interest was part of a new general direction taken by Institute in Via Panisperna, where for a few years there had already been talk of abandoning, albeit gradually, atomic physics, a field in which everyone had worked on for several years, and to focus research efforts principally on the problems of nuclear physics.

Towards the end of January 1932, *Comptes Rendus* issues began arriving with the famous notes of F. Joliot and I. Curie on penetrating radiation discovered by Bothe and Becker. The first of these described how penetrating radiation, emitted from beryllium under the action of alpha particles emitted by polonium, could transfer about five million electronvolts of kinetic energy to protons present in fine layers of various hydrogenated materials (such as water or cellophane). To interpret these observations, Joliot-Curie had initially suggested that it was a phenomenon analogous to the Compton effect... Soon after, however, they suggested that the observed effect was due to a new type of interaction between gamma rays and protons, different to that which occurs in the Compton effect.

When Ettore read these notes, he said, shaking his head: "They did not understand anything, it is probably due to recoiling protons produced by a heavy *neutral* particle". A few days later, the issue of *Nature* arrived in Rome containing the Letter to the Editor submitted by J. Chadwick on 17th February 1932, which proved the existence of the neutron on the basis of a classic series of experiments...

Soon after Chadwick's discovery, several authors realized that neutrons had to be one of the constituents of nuclei and began to offer various models consisting of alpha particles, electrons, and neutrons. The first to *publish* that the nucleus only consisted of protons and neutrons was probably D.D. Ivanenko ... But it is certain that, before Easter of that year, Ettore Majorana had produced a theory for light nuclei assuming that protons and neutrons (or "neutral protons" as he was calling them, at that time) were the only constituents, and that the former interacted with the latter only through quantum forces originated by the exchange of spatial coordinates (and not spin), if you wanted the system saturated with respect to the binding energy to be the alpha particle and not the deuteron.

He mentioned his draft of a theory to friends at the Institute and to Fermi, who was immediately interested and advised him to publish his results as soon as possible, even if partial. But Ettore was not interested because he judged his work to be incomplete. So Fermi, who had been invited to attend the Conference of Physics that was to take place in July of that year in Paris, in the broader context of the Fifth International Conference on Electricity, and who had chosen the properties of the atomic nucleus as his topic discussion, asked

Majorana permission to mention his ideas on nuclear forces. Majorana's reply was to forbid Fermi to discuss them or, if he absolutely had to, on the condition that he said they were the ideas of a well-known professor of electrical engineering, who was actually going to be present at the Paris Conference, and who Majorana regarded as a living example of how scientific research should not be conducted.

So it was on 7th July in Paris that Fermi gave his report on "The current status of the physics of the atomic nucleus" without mentioning the kind of forces that were later called "Majorana forces" and that essentially had already been conceived, albeit roughly, several months earlier.

In the *Zeitschrift für Physik* file dated 19th July 1932, appeared Heisenberg's first work on "Heisenberg exchange" forces, i.e. forces that involve the exchange of both spatial and *spin* coordinates. This work had a large impact on the scientific community: it was the first attempt at a theory of the nucleus that, however incomplete and imperfect, overcame some of the principle difficulties, which until then had seemed insurmountable. At the Physics Institute of the University of Rome, everyone was extremely interested and full of admiration for Heisenberg's results, but at the same time disappointed that Majorana not only had not published anything, but also did not want Fermi to speak of his ideas in an international congress.

Fermi strove again to convince Majorana to publish something, but every effort of Fermi and his friends and colleagues was in vain. Ettore sustained that Heisenberg had already said everything that could be said on the matter and that, indeed, he had probably said too much. Eventually, however, Fermi was able to convince him to go abroad, first to Leipzig (where Heisenberg worked) and then to Copenhagen, and he had the National Research Council assign him a grant for the journey, which began at the end of January 1933 and lasted for six or seven months.[2]

His aversion to publish or otherwise reveal his findings seemed, from this episode, to be part of his general attitude. At times, while talking to a colleague, he would casually state that the previous evening he had performed a calculation or developed a theory concerning some unclear phenomenon that had captured his attention or the attention of any of us at that time. In the discussion that followed, Ettore would at some point laconically pull out a packet of Macedonia cigarettes from his pocket (he was a heavy smoker) on which he had written in tiny but neat handwriting the main formulas of his theory or a table of numerical results. He would copy some of the results on the chalkboard, as much as was necessary to clarify the problem, then end the discussion, smoke the last cigarette, crumple the packet in his hand, and throw it into the waste basket.

---

[2] It lasted, as we now know, from 19th January to 5th August (1933), apart from the interruption from 12th April to 5th May (*Ed.*).

4. *His Trip Abroad*

In the winter of 1932–1933, Eugene Feenberg came to Rome from Harvard University under a *traveling scholarship* for "graduate students" of the University, with whom he spent three months in Rome and one in Leipzig. His stay in Europe was interrupted by the sudden invitation to return to the United States by the authorities of Harvard University, concerned about the political situation that was developing in Germany: in a few short months Hitler had managed to suppress civil rights and democratic freedoms and gain definitive control.[3]

During his time in Europe, Feenberg wrote his Ph.D. thesis on the scattering of electrons by neutral atoms; a work which established the "optical theorem" among other things, without really understanding its relevance and scope.

---

[3] The speed with which the events of 1933 in Germany came to pass is really amazing. On 30th January, at the end of a long government crisis, Hindenburg constitutionally appointed Hitler Chancellor of the Reich; on 27th February, the Reichstag fire prepared by the Nazis took place, and then subsequently blamed the Communists. On 28th February, taking advantage of the effect of the Reichstag fire, Hitler had Hindenburg sign a document which suppressed the clauses in the constitution that guaranteed individual and civil freedom. On 5th March, new elections were held and on the 21st of the same month the first meeting of the new Reichstag of the Third Reich took place; the Nazis did not yet have the majority, but were sufficiently powerful to succeed in delivering definitive power to Hitler over the following days (*Note of E. Amaldi*).

Feenberg and Majorana built an immediate rapport, but failed to establish close working relations, given that neither of them was able to speak the other's language; Feenberg had bought a small English–Italian glossary which he tried to use, but with modest results notwithstanding his eagerness and perseverance. They would therefore set themselves up in the same room of the Institute of Via Panisperna library, studying at the same table but only rarely, between one reading and another of the pages of recent publications, communicating with each other to show some formula written on a piece of paper.

Before leaving for Leipzig, Majorana published another paper, the one on the relativistic theory of particles with arbitrary intrinsic moment. It is his first work involving elementary particles (and not aggregates of particles like atoms or nuclei), and we will therefore discuss it shortly.

In January, Majorana left for Leipzig... which in those years was one of the major centers of modern physics; a group of exceptional young people had gathered around W. Heisenberg, including F. Bloch, F. Hund, R. Peierls, and, among the guests, E. Feenberg, R.D. Inglis, and E.G. Uhlenbeck. Feenberg remembers having attended a seminar by Heisenberg on nuclear forces, in which Heisenberg also spoke of the contribution made by Majorana to this topic: he said that the author was present and invited him to say something about his ideas, but Ettore refused to speak up. Leaving the seminar, Uhlenbeck expressed his admiration to Feenberg regarding the acuteness of the considerations made by Majorana, which Heisenberg reported.

Majorana in that period was very tied to Heisenberg, for whom he always maintained a deep sense of admiration and friendship. It was Heisenberg who easily convinced him, with only the weight of his authority, to publish his work on the theory of the nucleus, which appeared in the same year in both *Zeitschrift für Physik* and in *Ricerca Scientifica.* Heisenberg became aware of Majorana's remarkable research qualities, but also of the difficulties he always had in establishing relationships with people he had only recently met and, in general, with the outside world.

In Copenhagen, if not the major certainly one of the major physics hubs of the time, Ettore met Niels Bohr, C. Møller, L. Rosenfeld and many others. Placzek was also in Copenhagen during that period, and Majorana stuck with him as he had already known him for a few years.

In July, the Majorana family took a car trip to visit Ettore in Leipzig.

In the period he spent abroad, Majorana was struck by the economic and organizational capacity of the Germans, and he developed a great admiration for Germany which he expressed on a few occasions, particularly in a letter to Emilio Segré in which he tried to give an explanation — unacceptable to most of his friends — for the policies of the German government of the time.

When he returned to Rome in the autumn of 1933, Ettore was unwell due to gastritis which first appeared while he was in Germany. The origin of this illness is not clear, but family doctors linked it with the onset of a nervous breakdown. He began to frequent the Institute of Via Panisperna only occasionally and, as the months passed, not at all: he spent more and more time at home immersed in his study for extraordinary amounts of hours on end.

Besides physics, at that time he was interested in political economy, the fleets of several countries and their power relations, and the design characteristics of the ships. At the same time, his philosophical interests, which had always been alive in him, increased significantly, so as to induce him to ponder the works of various philosophers deeply, especially those of Schopenhauer. It was probably around that period that he produced the manuscript on the value of statistical laws in physics and social sciences, which was found among his papers by his brother Luciano and published (n. 10) after his disappearance by Giovanni Gentile junior.

To these old and new interests he added another, medicine, a topic that was perhaps born of his desire to understand the symptoms and the extent of his illness.

Many attempts made by Giovanni Gentile jr., Emilio Segré, and myself to bring him back to his normal life were to no avail. I remember that in 1936 he rarely left his home, not even to go to the barber; his hair had grown so abnormally that, following a visit to his home, one of his friends sent him a barber, despite his protests. None of us were able to find out, however, if he was still performing research in theoretical physics; I believe he was, but I do not have any proof.[4]

5. *His Appointment as Professor of Theoretical Physics and His Work in Elementary Particle Physics*

In the meantime, several other young people began maturing in the field of theoretical physics: Gian Carlo Wick, graduated from the University of Turin with Somigliana, had come to Rome after a period spent in Göttingen and Leipzig; Giulio Racah, graduated in Florence with Enrico Persico, always divided his time between Florence, Rome and Zurich, where he worked under the guidance of Pauli; Giovanni Gentile junior, of whom we have already spoken; Leo Pincherle who

[4] In fact, regarding this period we now know (at least) of the replies Ettore wrote to his uncle Quirino [distinguished experimental physicist], in which Ettore patiently forms, at the request of his uncle, the theories designed to explain Quirino Majorana's various experiments. From these letters it is also — more importantly — evident that in those years Ettore did continue to perform original theoretical research in physics [for instance, in quantum electrodynamics] for his own purposes. (Ed.)

had studied at Bologna before coming to Rome, and Gleb Wataghin, who emigrated from Russia to Italy and studied in Turin, where he taught and worked for years.

The time had come for a new contest in Theoretical Physics; the first and only competition for professorships in this subject took place in 1926, which resulted in Enrico Fermi's placement in Rome and Enrico Persico's in Florence.[5] The new competition was tendered at the beginning of 1937 at the request of the University of Palermo, driven to do so by Segré, who in the meantime had become Professor of Experimental Physics at the same University.

There was of course the problem of making Ettore apply, who seemed to prefer not to be involved and who in any case had not published any works in physics for some years. Fermi and various friends persevered and finally persuaded Majorana to take part in the contest, who thus published his famous work on the symmetry of the electron and the positron in *Nuovo Cimento*. The Selection Committee for the professorship competition for Theoretical Physics at the University of Palermo was appointed by the Minister of National Education, as prescribed by the then fascist laws, with the following members: Antonio Carrelli, Enrico Fermi, Orazio Lazzarino, Enrico Persico, and Giovanni Polvani. The competitors were the five mentioned above as well as Ettore Majorana. The

[5] In addition to Aldo Pontremoli in Milan. (Ed.)

Commettee held a primary meeting during the month of October 1937; but was soon invited by the Minister to suspend activity[6] — in order to be able to proceed with the appointment (based on Art. 8 of Royal Decree n. 1071, 20th June 1935) of the competitor Majorana as Professor of Theoretical Physics at the Royal University of Naples. The above article referenced special merit; and it was introduced a few years earlier in order to allow the appointment without competition of Guglielmo Marconi to the Chair of Electromagnetic Waves at the University of Rome.

His major scientific contribution came from his last three works. The first of these, the origin of which I have already mentioned, followed the famous article by Heisenberg…(*omissis*).[7]

Work n. 7 on the relativistic theory of particles with arbitrary intrinsic moment…(*omissis*).[8]

In his final work, regarding the symmetry of the electron and the positron, Majorana began with the observation that the relativistic theory of Dirac, which had led to the prediction of the existence of the positron, shortly after confirmed by experiments, hinged on the Dirac equation which is completely symmetric with respect to the sign of the charge: but that symmetry was partially lost in the subsequent development of

---

[6] See Chapter 2. (Ed.)

[7] For the more technical aspects see — in addition to Chapter V here — the original text of E. Amaldi: ref. Footnote 1, and the Biography. (Ed.)

the theory, which described the void as a situation in which all the negative energy states were occupied and all positive energy states were free. Exciting an electron from one of the negative energy states to one of the positive energy states, left a gap with positive energy, which could be interpreted as the anti-electron [*or positron*]...This asymmetric scenario also results in the need to cancel, without any sound justifying principle, certain infinite constants, such as charge density, due to the negative energy states. Based on these observations, Majorana developed a theory in which a neutral particle, say, the neutrino, is identified with its anti-particle, the anti-neutrino...(*omissis*).[8]

Appointed professor of theoretical physics at Naples in November 1937, Ettore Majorana moved to that city in early January of the following year. In Naples, he befriended Antonio Carrelli, Professor of Experimental Physics and Director of the Institute of Physics of the University.

In Naples, as he had always done in Rome, he led a very solitary existence; he went to the Institute in the morning when he had lectures to give, and in the late afternoons he took long walks around the livelier neighborhoods of the city. He fulfilled, as he had always done in all of his duties in the past, his teaching obligations with great care and commitment. The manuscript of his lectures on quantum mechanics shows how his teachings were conducted in a manner very similar to present day approaches.

In Naples, as during the previous years in Rome, Majorana was tormented by his illness, which inevitably influenced his mood and his personality. This perhaps explains the excessive sorrow that Carelli says he felt when, after a few months of teaching, he realized that very few of the students were able to follow and appreciate his always extremely high-level lectures.

On 26th March 1938, Carrelli is extremely surprised to receive a flash telegram[8] from Ettore Majorana in Palermo, telling him not to worry about what was written in the letter that he had sent. Carrelli awaited the arrival of the letter written in Palermo a few hours before the telegram was sent[9]; Ettore Majorana wrote it with great calm and just as much decisiveness, finding life in general, and his in particular, absolutely useless[10] and therefore deciding to end it all. The letter

---

[8] In fact, an urgent telegram, as stated by Carrelli in document D/ME6, Chapter 10. See Chapter 1. (*Ed.*)

[9] As we know from Chapter I, the letters were actually two (MC/N and *MC/P*, Chapter 9). See Chapter 1. (*Ed.*)

[10] In 1966, Amaldi did not yet know of the correspondence that we subsequently rediscovered. Until its discovery in 1972, therefore, the letters of Majorana were based on memory and "hearsay". See instead the Letters *MC/N* and *MC/P*, Chapter 9. (*Ed.*)

was unfortunately lost,[11] but one particular phrase was etched in the memory of Carrelli and it read something like this: Do not take me for an "Ibsenian girl", the problem here is much greater...The letter ended with a warm goodbye to Carrelli, thanking him for the friendship had shown in preceding months.

Carrelli, upset by this letter, immediately phoned Fermi in Rome, who contacted Ettore's brother Luciano: he[12] immediately went to Naples and began a frantic search for information on Ettore. This search, conducted both in Palermo and Naples, established that Ettore had departed from Naples to Palermo on the Società Tirrenia steamer on the night of 23rd to 24th March [*actually, it was the evening of 25th March (Ed.)*] and arrived in Palermo where he stayed a couple of days and then, on the 25th [*actually, the morning of the 26th (Ed.)*], he sent both the letter and the telegram to Carrelli. On the evening of the same day, he boarded the steamer for Naples. Prof. Michele [*actually, Vittorio (ed.)*] Strazzeri of the University of Palermo saw him that night on board and even at dawn, while the steamer entered the Gulf of Naples, saw him sleeping in his cabin. A sailor testified to seeing him at the stern of the ship after Capri, not

---

[11] It was actually immediately taken into possession by the family (probably on 29th March by his brother Salvatore), and thus remained lost until our discovery in 1972. (Ed.)

[12] Documents *D/ME7* and *D/ME6* (Chapter 10) show that the brother was Salvatore. In reality, both brothers, Salvatore and Luciano, immediately went to Naples. See Chapter 1. (Ed.)

long before it docked at the pier in Naples. According to the Società Tirrenia Naples office, Majorana's Palermo–Napoli tickets were found among those delivered among the arrivals at Naples, but this was never confirmed.

Investigations continued for over three months by both the Police and the Caribinieri, with the personal interest of Mussolini, whom his mother had turned to. The family promised a then very conspicuous reward of 30,000 lira to anyone who forwarded information regarding Ettore, and published appeals for him to return home in major newspapers for months; the Vatican sought to determine if he had closed himself in a convent. But all attempts were in vain. No trace was ever found: all that was learned was that, a few days before Ettore Majorana departed for Palermo, a very troubled young man whose somatic and psychic characteristics seemed to relatives to correspond with those of Ettore presented himself at the Church of the Gesù Nuovo, located in Naples close to the Hotel Bologna, where he lived.[13] In addition, Father De Francesco, former Provincial of the Jesuits, who had received the young man, seemed to recognize the photograph of Ettore shown to him by relatives. The young man enquired with Father De Francesco to "experiment with religious life", a phrase which, *according to the brothers* should be understood as "performing spiritual exercises". In fact,

---

[13] And even closer to Via Tari, which was the seat of the Institute of Physics. (*Ed.*)

they do not believe that his intention was to express the desire to undertake a religious vocation, but a simple desire to retire in meditation.[14] The answer was that, yes, he could be hosted, but only in the short term — as a definitive solution would have required his initiation to the Order as a Novitiate —: the young man replied "Thank you, excuse me," and left.

The hypothesis believed most probable among friends was that he threw himself into the sea, but all the experts of the waters in the Gulf of Naples argue that the sea, sooner or later, would have returned the remains.

Only 30 years later, someone who he had never met or had known only very superficially, conjured the notion of a kidnapping or an escape in relation to a hypothetical episode of atomic espionage. But for those who lived in this era of nuclear physicists and knew Ettore Majorana, such an hypothesis is not only devoid of any foundation, but it is absurd both from a historical and human perspective. A few years after his disappearance, in a discussion on the matter with common friends, Fermi observed that, with his intelligence, if Majorana had decided to disappear or ensure his body wasn't found, he would certainly have succeeded.

---

[14] The interpretation of "the brothers" given by E. Amaldi does not seem to agree with the information that followed. Furthermore (if the "young man" was, as it seems, Ettore Majorana) it should not be forgotten that, for a physicist, the word "experiment" will have well-defined implications which cannot be reduced to "doing exercises". (*Ed.*)

Nothing else was ever found: everyone was left with a sense of deep sorrow for the loss, of a relative, or of a friend, kind, shy, and withholding external manifestations, so evidently affectionate even if deeply bitter: a sense of frustration for all what his genius did not left behind but that could have still produced, were it not for his absurd disappearance; and above all, a deep sense of atonishment and admiration for his figure of a man and a thinker that had passed among us so quickly, like one of Pirandello's characters loaded with problems that he bore with him, alone; a man who had been able to so admirably answer certain questions of nature, but who had sought in vain to find a reason for life, his life, even if it was for him far more full of promise than it is for the vast majority of men.

Edoardo Amaldi